MOVING TO ITALY
A RELOCATION ROLLERCOASTER

KDP
2024

Version 1.0

Copyright © 2016 by Stef Smulders
No part of this publication may be reproduced, distributed or transmitted in any form or by any means, including photocopying, recording, or other electronic or mechanical methods, without the prior written permission of the author, except in the case of brief quotations embodied in critical reviews and certain other noncommercial uses permitted by copyright law.

Disclaimer
This is a work of fiction. Names, characters, businesses, places, events and incidents are either the products of the author's imagination or used in a fictitious manner. Any resemblance to actual persons, living or dead, or actual events is purely coincidental.

Stef Smulders
Frazione Cassinassa
27040 Castana (PV), Italy
www.stefsmulders.nl

Moving to Italy/Stef Smulders
ISBN 9798321585436

MOVING TO ITALY
A RELOCATION ROLLERCOASTER

Stef Smulders

Emese Mayhew
(translation)

K D P
2024

For Nico
My own padrone di fiducia

It once occurred to me that one way to talk about Italy would be simply to make a list of all those Italian words that are untranslatable, or whose translation tells you next to nothing, and then give dozens of anecdotes showing how they are used.

An Italian Education - Tim Parks

November 2008

When we bought our home nine months ago it was ready to move into. And now?

We are shipwrecked in the kitchen of the downstairs apartment. A single sheet of plastic between the hall and the sitting room is the only thing that protects us from the heavy dust of the building site. All day, we are assaulted by the sound of workmen shouting, drilling and hammering. A couple of hours ago the electricity cut out and it's starting to get chilly in here. Every evening we escape upstairs via the dusty, grimy staircase, where we try to find solace by watching TV in our future living room. The living room is also separated by a sheet of plastic from the kitchen, the bedroom and the office. There are gaping holes in the walls in all of these three rooms, made weeks ago in preparation for the doors and a new window. Now they are serving as tunnels bringing in the draught and the cold. Exhausted and numbed from the endless turmoil surrounding us, we are staring out into space in silence. We are hardly aware of what's on the screen.

WHAT HAVE WE LET OURSELVES IN FOR?

I

Pavia

SEPTEMBER 2007 – FEBRUARY 2008

Non ci sono problemi

With my right foot still on the pavement, the estate agent's car was already pulling away. My reaction was fast: I pulled both legs inside and slammed the car door, averting an accident. The estate agent obviously had no time to waste! We were going to look at two properties in the Oltrepò Pavese, the area lying south of the river Po, which traverses Northern Italy. I sat in the front and the estate agent prattled on in hundred-mile-an-hour Italian. I only understood bits of what he was saying, partly because I was too disconcerted by the traffic which we were navigating with Italian flair.

For the last few weeks, we had lived in the quiet, historical, university town of Pavia. In the next 6 months, I was going to continue with my MA in Medieval Culture, and my husband, Nico would enjoy his well-earned sabbatical. He was going to hoover, do the shopping and cook, whilst I could immerse myself in times gone by. But there was this secret, unspoken wish that didn't leave us alone: could we...., what if we..., imagine if...?

And already, just a couple of weeks into our stay in Pavia, we started looking at properties, with the intention of permanently settling down and setting up a B&B! Soon after our arrival in Pavia, we discovered the wine region of Oltrepò Pavese, an area about half an hour's drive to the south of Pavia. It was love at first sight. What beautiful countryside! And this is how our secret wish began to take shape: to find our own idyllic home on the top of a hill with panoramic views! In one of the free leaflets from the numerous estate agencies (*agenzie immobiliari*), our excited eyes spotted the perfect house that ticked all our boxes. We were now on our way to this house, with an estate agent whose main talents seemed to be smooth talking and rally driving.

Once we got out of Pavia, the roads became quieter, and I was able to follow Olita's - as he was called - Italian a bit better. He was busy showing off his property know-how and reassuring us about the top quality of the houses we were about to see. If there was anything not to our liking, it could be easily sorted, without any additional costs, he said. He had already made an agreement with the owners. *"Non ci sono problemi!"* he exclaimed with much enthusiasm. If we didn't like the colour of the house, it could be painted over, before completion, in any colour at all, even violet, maintained Olita. *"Non ci sono problemi!"* And the garden that had become a jungle from months (probably years?) of neglect would be completely cleared out, just for us.

We took in the landscape in front of us: it was mainly flat, covered in rice fields (growing the famous Italian *risotto*), farmland and poplar plantations, as far as the eye could see. Along the country road, we were driving past settlements: an endless mish-mash of houses and farm buildings of all shapes and sizes. We raced through small villages with stores, restaurants, and cafés. Olita was consistently indifferent to the numerous white traffic signs warning of upcoming speed cameras. Did his employer pay the fines? Or was it going to become a hidden charge on our bill? We were fully aware that we were going to have to pay Olita commission if we were to buy our house through him. We had done our homework in the Netherlands and were well-prepared for all the traps that a would-be house buyer could fall into when trying to buy a house in Italy. We were on high alert! Olita, unaware of my misgivings, drove on at full speed. Here and there along the side of the road, there were small shrines erected by friends and relatives of beloved maniacs, who had died in tragic road accidents. Olita didn't seem to worry about suffering the same fate; he overtook slow drivers without mercy, regardless of whether the white line was broken or solid. Later, having lived in the Oltrepò for several months, we discovered a *santuario* nearby; a memorial chapel for all the victims killed in road accidents in the area. The legendary recklessness of Italian drivers might have some foundation after all. Olita, for his part, did his utmost to conform to the stereotype. Occasionally, we met two cars side-by-side coming from the other direction, but luckily three cars in a row could easily be accommodated on this two-lane road. *Non ci sono problemi.*

We reached Ponte della Becca, the one-kilometre-long iron bridge built in 1912 that spans the merging of the Po and the Ticino. The Oltrepò stretched on the other side, flat at first, but soon undulating with hills. There in the distance our dream house was waiting for us somewhere. We saw the first vineyards appearing here and there. On one of the hillsides, we spotted a remarkable-looking castle and we inquired about it from our local regional expert, a.k.a. Olita. "Which castle is that?" we asked full of curiosity. He didn't know. But *"Non ci sono problemi,"* he would investigate and let us know. Maybe our house was not going to be violet after all.

It soon became apparent why Olita was in such a hurry: he was lost and was zooming up and down the hills in search of familiar landmarks. Against all expectations, we managed to find our chosen house, which didn't look as perfect as we at first had thought, not even if Olita would have it painted violet. On one side it leant against a slope, and the other side was blocked from view by an unsightly shed. The garden was no bigger than a postage stamp. What a shame. Luckily, on the advice of Olita's *Agenzia*, we had also made an appointment to view another property that was on offer at a bargain price. This second house didn't look appealing in the brochure: a faded grey concrete block without any character. But now that we were here...we might as well take a look.

It took Olita a lot of cursing and muttering under his breath during the second stretch of our mystery tour, to finally bring us to the cheaper property. The frontage made no false promises. There were not enough colours in the rainbow to change that. But the inside! The house was made up of two apartments, each a hundred square metres. The downstairs apartment was completely modernised, had brand new flooring, central heating, a fitted kitchen, and there was a sitting room *with* sofas and a ready-to-go modern bathroom. The apartment was ready to move into as soon as gas and electricity were connected. We felt enthusiastic.

After having seen the downstairs flat, Olita led us upstairs and opened the shutters of the bedroom overlooking the valley. An enchanting view of rolling hills and vineyards in the style of impressionist paintings unravelled before our eyes. In the distance, we recognised the characteristic but still enigmatic castle from earlier. And a bit further on, there was *another* castle. And over there *another* one. We were sold. *Non ci sono problemi!* For once we all agreed!

Via Moruzzi

Our base in Pavia, which we were renting until we found a house to buy, was a flat owned by Giorgio and Franco. It was a lucky find. In the summer of 2007, we visited Pavia for a week to find an apartment for my six months study abroad and Nico's sabbatical. At first, that week seemed as if it would end in total failure because all the suitable apartments we found on the Internet in the Netherlands fell by the wayside one by one. In one case, for example, we were allowed to view our chosen flat but later it transpired that it wasn't quite clear whether the present tenants were really going to leave. Why didn't the owner tell us this earlier, we wondered feeling annoyed. What was the point in looking at a flat that wasn't (yet) available? Did the owner worry that he would disappoint us and let us carry on with the viewing? But now we were even more disappointed. Maybe this is the Italian way of doing things, we thought, quite put out by the way things were handled.

We had nothing left but to hope that the last of the apartments we had selected was still available and that we would like it. Although our appointment for the viewing was later on, in the evening, we decided to have a quick look around the neighbourhood in daylight. We saw at the entrance, where the doorbells and tenants' names were listed next to the apartment numbers, that the name plate next to our chosen apartment was empty. The flat was seemingly still free: that was at least a positive sign! We returned that evening, full of expectations, and rang the bell. But what on earth was that? We stood looking in disbelief at a name next to the number of *our* apartment! That could only mean one thing, we concluded crestfallen: the flat had been rented out today. But surely the owners wouldn't let us make a wasted journey? Did we check properly this morning? Was it just the name of the previous tenant? We hoped for the best and pressed the button again.

The gate buzzed open and we entered with apprehension. The apartment door was opened by a young couple, with deadpan faces. They showed us around the whole apartment, explained its pros and cons and provided other useful information. It turned out to be a quite sparsely furnished, minimalistic and not too spacious dwelling, but because we had no other alternatives, we offered to rent the place at the end of the viewing. "Yeah," said the girl a bit sheepishly, "there is a little problem." The flat was indeed already rented out. This crucial bit of news had a devastating effect on us. What were we supposed to do now? We would never have enough time in the remainder of the week to find another place. The girl saw our desperation and took it to heart. Suddenly she remembered a friend who had a furnished apartment that he might be prepared to rent out. "Yes please, we are very interested," we both shouted, clutching at straws. So she rang her friend, Giorgio, who agreed to meet us at Pavia train station and take us to his flat on Via Moruzzi.

Arriving at the station we couldn't see any Italian who looked like they were there to meet someone. We decided to wait at the entrance. Before long my mobile was ringing. "*Sono qui,* I am here," I heard a voice say, and at the same time saw a man approaching us: that must be Giorgio. He had been observing us from a distance to decide whether we were *persone serie*, serious people. Luckily he must have thought so and soon we were driving up behind him to the flat that was going be our salvation. To our great relief, his parents' flat (because that's what it was), was by far the best of all the accommodation we had viewed. Our trip was a success after all, not thanks to our careful preparations, but because of the quick thinking of an Italian, who knew someone, who... Was this a taster of our forthcoming experiences in Italy?

Vista sui tetti di Pavia

La Nagel: with her oversized sunglasses (dark round lenses surrounded by thick plastic frames), which she had just unearthed from the depths of her handbag to protect her eyes from the strong Italian sun, she bore a strong resemblance to Sophia Loren in her heyday. She was a researcher in Medieval Astrology and my future collaborator at the University of Pavia. Her straight hair, dyed raven black, gave the impression of an eventful past, a girl who must have turned many heads in her day. But today, the staircases of the old university buildings demanded her every last breath and she did her utmost to avoid the characteristic cobblestones of Pavia's historical streets: her fashionable shoes and tired feet couldn't even contemplate walking over them.

She lived in Milan, as did nearly all my other colleagues in the faculty, and commuted every day by train to Pavia. The journey was too dangerous by car because in the autumn the plain of the River Po is shrouded in a persistent, thick fog that can last for days. The sub-faculty, Medieval Philosophy was led by *la professoressa* Crisciani and was made up of five researchers, all of them women. Last year, I had succeeded in convincing *la professoressa* that her research group would be the perfect setting for my placement. But when I turned up last summer to visit the group for the first time, they could barely hide their astonishment. They expected the intern to be a woman. The fact that on my profile picture which I e-mailed to them I was obviously bald and was sporting a beard, was apparently not enough evidence to prove my masculinity.

General common sense does not seem to apply to Medieval Philosophy!

I received a warm welcome, nevertheless, and my arrival was celebrated with lunch at a restaurant in Pavia's city centre, called the Osteria alle Carceri, the dungeon inn so to speak. Hmmm, could I detect a hint of foreboding in this name? Following *la dottoressa* Nagel's advice, I ordered a *risotto bianco*, which promised to be delicious. But to me, the *risotto* seemed only to consist of rice, butter and cheese without any further ingredients, it tasted rather plain and bland. To the unexpected question of whether I liked it, I of course answered *"buono"* in order to avoid antagonising my medieval friends at such an early stage. Luckily, some time later, completely out of the blue, Giorgio forbade me ever to visit this very restaurant, as it was well-known for its over-pretentious food!

After lunch, *la professoressa* made a quick exit. She was not heading to the university; instead she was going home to look after a sick elderly relative who had suffered a stroke recently. My *professoressa* was very sorry to say goodbye to me so soon, but she was certain that we would come across each other regularly in the next couple of months. *La* Nagel was in charge now to give me a tour of the centuries-old university. She showed me the university buildings, the anatomy room and the library. At one point the conversation turned to where I should stay in Pavia during my 6 month-long visit. My *dottoressa* hadn't the faintest idea how much trouble it had caused us in the last couple of days just to secure a roof over our heads. Her advice was well-meant albeit naive: *"Dovreste prendere un appartamento con la vista sui tetti di Pavia*! You ought to hire a nice apartment with a view across Pavia's rooftops!"

Persone serie

This was the last straw! Giorgio was burning with rage because of his brother's Franco's last comment, made in jest: "*Siete quasi clandestini*! You are some sort of illegal immigrants!" How could he say something like that, how could he act so *maleducato*, blunt, towards such respectable people as we were in Giorgio's eyes. *Persone serie, persone brave.* Because of the way Giorgio emphasised that last bit, we got the impression that he didn't come across many people like that in Italy. Is Italy full of untrustworthy characters who cannot be taken seriously? Who say one thing and do another? We would soon find out. Luckily, according to Giorgio, we didn't belong in that category.

Although they were brothers, Giorgio and Franco had strikingly different personalities. Giorgio was short and squat like a rugby player with dark wiry curls; he had a beard and wore glasses and everything he said seemed to have been well thought out. He often had an introspective air about him. Franco, on the other hand, was tall and slim, with thinning hair, and had no beard or glasses (the latter for reasons that would become apparent later). Franco moreover, had a nervous energy that didn't let him sit still, paired with impulsive tendencies: he blurted everything out directly whilst looking straight at you as if waiting to see your reaction. Each brother seemed to impersonate a different aspect of '*the* Italian': Franco, the jovial, carefree, cheerful, not-to-be-trusted Italian of the proverbs, as most outsiders imagine them; Giorgio, the caring, pessimistic and slightly depressed version of the Italian, the kind you come across in Italy quite often. It's not for no reason that many Italians will answer 'how are you?' with *"non c'è male"* 'not too bad' instead of with *"bene,* very well". Franco always greeted everyone with a deafening *"Tutto bene?"* He meant this as a rhetorical question because he repeated it every time you fell into a momentary silence: *"Tutto bene*?" He never really listened. Giorgio, on the other hand, often engaged you in deep and serious conversations about the shortcomings of Italy and its people and about the bleakness of his own prospects. Like every coin, Italy seems to have two sides: manic and depressed.

The exchange intensified between these brothers representing the extreme polar opposites and (we felt) it was growing into a full-blown argument. We understood very little of what was said, we picked out the words "Schengen" (pronounced: shyenghen) and *"Sei pazzo*! You are crazy!" Disagreement? Oh well, this was just the typical way feelings were expressed, in keeping with the Italian temperament. A good example of 'much ado about nothing'. When the dispute was finally over, Giorgio carried on irritably with the complicated and extensive paperwork that the anti-terrorism legislation required him to fill in. We were renting his apartment as foreigners with temporary residence permits and the Italian government needed to know all the ins and outs.

Giorgio's and Franco's flat forms part of a so-called *condominio*, an apartment complex. These can be found all over the small town suburbs in Northern Italy: 3-4 storey buildings, surrounded by a garden, with their own car park and protected by a metal railing. The gate securing the area surrounding a *condominio* (safety first!), is not just an ordinary one, but a *cancello a telecomando*, a remote controlled gate! And it's also fitted with a flashing light because a house or a *condominio* without such a gate and orange light is like a monarch without a crown. You have only really made it in life if you successfully moved into a house equipped with *both* an automatic, remote controlled gate *and* an orange flashing light. There were also supposed to be little warning signs to prevent accidentally trapping children completely automatically between the wall and the gates and squashing them into French fries when opening the gate. Safety first.

The flats in Giorgio's and Franco's *condominio* were accessible through a shared lobby. There were no external corridors. The basement consisted of small box rooms and garages. The management of the *condominio* was carried out by an unavoidable group of owners, the 'neighbourhood watch' who (safety first) ensured cleanliness, peace and routine. The *condominio* was situated in Via Moruzzi, west of the city centre and Pavia's railway station. It was surrounded by a beautiful garden and there were plenty of covered parking spaces reserved for each flat. The wall in the brand new lobby was clad in polished natural stone. And of course we received our own genuine *telecomando* for the gate, which was naturally equipped with a lovely flashing orange beacon. But first, we had to be cleared of any suspicion of subversive intentions that could possibly link us to terrorists. Giorgio did his best to arrange this for us, but the pile of paperwork full of official jargon made it a nearly unbearable chore.

Whilst Giorgio was focused on deciphering the instructions, Franco, completely unaffected by the previous argument, started up a friendly chat. About reading glasses and the dangers of wearing multifocal lenses, for example. Franco had heard stories from people wearing varifocals who fell down staircases because they couldn't see the steps properly. "Deadly!" he asserted. He was adamant not to wear glasses of that sort or, to think about it, of any sort, even though he was short-sighted. As a result, he read the year on our 1875 Bols Genever Gin bottle as 1575. Franco was preoccupied, just like nearly all Italians, with danger and health. We noticed this when he showed us around the neighbourhood, shortly after we had moved into the flat. He pointed out the hospital, the *farmacia*, the pharmacy and the headquarters of the Red Cross *and* the Green Cross, all these facilities available to us within our district. We as *persone serie* were completely safe, he seemed to say.

By now we had already spent a couple of weeks living in Giorgio and Franco's flat, who on this fine evening cleared us of any suspicion of terrorist activities. We had to drink a proper Dutch toast to that. Bols Genever from 15... no, wait, 1875.

La perizia

"The ceiling is two and a half metres high, you see," said Olita, the estate agent, in a self-assured tone. "No, it's two metres seventy," came the impassive correction from Luigi Buttini, our hired *geometra*. A *geometra* is a typical Italian professional, whose expertise encompasses everything from architectural engineer to planning specialist. He is virtually indispensable when buying and vetting a house. We hired Buttini to inspect the house in the Oltrepò which we set our heart on. We were already pretty taken by the house but we wanted to avoid ending up with a fool's bargain. The fact that we couldn't trust our estate agent Olita in this respect had already become clear at our first viewing.

"Two metres fifty," countered Olita, annoyed and abrupt because of Buttini's correction. "Let's measure it," concluded Buttini, sure of himself and equipped with all the necessary tools to make this possible. The result of this little duel of masculine egos was that the height was established as two metres seventy-five, meaning our *geometra* won. We suppressed a smile. Both men had been trying to get on top for some time now, Olita always on the alert for any mistakes Buttini could be caught making.

Buttini checked everything: did all the measurements tally with those in the land registry? Had anything been modified or extended illegally? Was the size of the plot of land correct? *"È tutto in ordine, non ci sono problemi,"* Olita shouted out time and again, offended that we brought in a real expert to check on him. But we were well prepared, and we bore firmly in mind all the disasters that could befall someone trying to buy a house in Italy. There was already something that didn't seem to be right: the piece of land that Olita's advert promised us was at least two thousand five hundred square metres. On our first visit, he showed us the borders of the land, which according to him extended to the end of the little brick building, called the *rustico*.

Back home after the viewing, in the middle of the night, awake with excitement over the fact that we had probably found our dream house, I suddenly realised that this couldn't be right. I thought that the amount of land around the house seemed to be too small (where was the swimming pool supposed to go?) and this could be a reason not to buy. But wait a minute, I thought: the house itself measures 11 by 11 metres, which is 121 square metres. The house should fit into the land over 20 times. But that was impossible on the piece of land that Olita had shown us.

Now that we had brought our own surveyor, this question should soon be resolved. The stocky figure of Buttini was wading through the tall weeds (an outstanding job for Olita?), stumbling across leftover roof tiles that had been thrown away haphazardly by roofing workmen. Olita was bounding along behind him like an overexcited puppy. Panting for breath, he called out one more time, warning us that the grounds beyond the *rustico* did not belong to the house and that we shouldn't be trespassing: it was *proprietà privata*! With slight panic in his tone, Olita shouted across to the owner to ask for his support. But the owner stood at the front of the house and didn't hear him. Buttini pushed on, entering illegal territory. Or maybe not? No, because he concluded that our piece of land stretched completely to the walls of the neighbouring house. The land registry documents confirmed this fact. Two nil to our *geometra!*

Olita was becoming ever more miffed and he had already started the day off in a bad mood. "You are late," he called out in annoyance. "I don't think so, we agreed half past nine," I said. "Nine o'clock!" he insisted. The owners had also had to wait half an hour, but they didn't hold it against us. "*È tutta colpa sua,*" said the lady of the house smiling at me. "It's all his fault." It was clear that they weren't on the best terms with Olita either. We could turn this to our advantage. I asked the woman if there was any other interest in the house at present. "There is *some* interest," she said, but she didn't sound convincing.

Olita rang us a couple of days later to ask us in an aggressive tone why we hadn't let him know yet whether we were going to make an offer on the house. Namely, everything was in order, we only needed to pay a deposit and sign a temporary sales agreement. But we - being well-prepared - had other ideas and we made this clear to him: "First we do a *perizia*, a survey, then we examine all the paperwork: the land registry, the ownership documents, outstanding debts, etc.. Then we will see how it goes." "Shouldn't we actually check if the neighbours would want to buy the land?" we asked our expert. According to the law, neighbour farmers have the first priority to buy when someone is selling farmland adjacent to their properties. "*No, no, non ci sono problemi,*" called out Olita immediately, but he was going to check just to be on the safe side: we were right.

In an hour or two, Buttini came to the conclusion that it all looked pretty good, and even better: a house as big as this for this asking price was a real *affare*, bargain. Now we only needed to go to the town hall in Montecalvo to make sure we would not stumble on any difficulties regarding the land-use plan, and then we could finally make our first offer on the house. We felt the tension rising. Could anything still go wrong?

Software potente

The telephone started ringing in our flat in Pavia. It was Giorgio, sounding rather sheepish. At first, he confessed, he had been in doubt whether to call us, but there was no alternative, because there was a problem. He found it really embarrassing having to do this, and he even considered not ringing us at all. In the end, he had decided that it was better to talk to us about this issue, but now he had doubts. "Come on, just spill the beans!" we insisted like Dutch foreigners, not appreciating the delicacy of this Italian embarrassment. Well, he had counted the first rent that we had paid, and it was a hundred euros less than expected. We were amused by all this hesitation from this timid Italian who had got himself in a fix because he didn't want to offend us and as a result he had nearly lost out on a hundred euros. We invited him to come over; we could pay the outstanding amount immediately. "No, no, it can wait, it's not a problem," he was evading our invitation. But we insisted on paying him now, so that we would avoid months of awkwardness and avoiding each other.

Giorgio arrived together with his brother Franco and decided to take this opportunity to connect us to the Internet. We had a telephone, an ADSL-router, and our laptop was ready. The only thing missing was the Alice software that the Italian internet provider *Telecom Italia* used. This publicly owned company is not famous for its user-friendly software, and rumours regularly surface of Telecom's bureaucracy, whispered about with barely veiled contempt in bars and cafés by unlucky victims. The first problem in our case was not with the software but with the electric cables: we needed an extension lead. We could construct a temporary set-up just for installation purposes, but in the long-term, we needed a permanent solution. *"Ce l'ho a casa,"* said Franco. "I have one at home." But his wise words fell on deaf ears. Giorgio and I were already completely lost in the Alice software, and we stopped responding to outside conversation. We quickly started to feel like we were trapped in Alice in *Wonderland*. It was such a mess! Franco repeated that he had an extension lead and he could go and get it. But, again, he got no reply.

"Per continuare si deve installare il nostro software potente," announced Alice happily. "Now you need to install our powerful software." I glanced at Giorgio with some disbelief. He understood immediately and said in an ironic tone: "Well? Do you want that powerful software from Telecom Italia on your PC?" *"È proprio la parola potente che mi fa paura,"* I grinned. "It's the *powerful* bit that I am worried about." But we had no other choice and with my eyes firmly shut I pressed '*Installazione*'. Beyond all expectations, everything went smoothly and the software got installed. In the meantime, Franco repeated two more times that he had an extension lead and he could go and get it. By this point he was pacing nervously up and down the room. "What's the matter?" asked Giorgio. "I can get the extension lead, give me the car keys, I will be back in a minute," answered Franco grumpily. Giorgio did as he was told.

Franco was away for over an hour with Giorgio's car, who was therefore stranded in our flat. Have a drink then, a bit of Bols, the genuine Dutch *grappa*. "No, no, maybe just a sip," said Giorgio. They became several sips whilst we were waiting for the extension lead. In the end, we heard Franco outside, parking the car. Downstairs the hall door slammed. He burst in in a fit of anger and frustration.

He couldn't find the lead anywhere. He had turned the whole shed upside down because it had to be there somewhere. A couple of days ago he had it in his hands. "And now it was nowhere to be seen, *porca miseria*!" We comforted him with a shot of special Dutch *grappa*. Nothing bad could befall us now: our software was *potente*!

Frazione Crocetta

We were sitting in the doctor's waiting room. A waiting room like any other, with rickety chairs, dried out pot plants and piles of fashion and gossip magazines, of the sort that you would never touch with a bargepole, but on occasions such as these, you leaf through them absent-mindedly in order to kill time. In this case, we didn't even know most of the 'celebrities'. It was still interesting to see how, in Italy too, they were photographed in all kinds of compromising and provocative situations and were subjected to speculative and juicy gossip. Having seen all the saucy stories in the tabloids, we had to come to the conclusion that the life of a celebrity is the same all over the world. The walls of the waiting room were covered in notices about the yearly flu jab and other information leaflets, like the one about the *mafia*: a fund that supported victims of the *mafia* asked for contributions from those present.

It was Friday afternoon and we were the only ones present in the waiting room because the GP only held surgeries on Tuesdays and Thursdays for the village of Montecalvo Versiggia (population 600). The waiting room of the GP was located in the town hall in Montecalvo and we were not waiting for him but for the county *architetta*, Roberta. She would advise us about planning permission, land-use restrictions and so forth, in relation to our planned acquisition of the house in *frazione*, hamlet, Spagna. On her decision our dream plan would stand or fall, so it was a very tense visit, just as tense as if we were visiting the doctor.

The town hall is in the 'centre' of Montecalvo Versiggia, in the *frazione* Crocetta, which is itself a small hamlet with a handful of residents. You won't find any high-rise, modern architecture here, only some more or less renovated houses with more or less neglected gardens. And in between the houses, nice views over the hills, covered in vineyards. The name Crocetta means little cross, or crossroads because the *frazione* occupies the junction of three roads. Once this was an important travel route and in the old days, they had an inn here, which has now become the restaurant La Verde Sosta. Chef Grazia and her husband Giuseppe (Pino to friends) are in charge of the restaurant. Giuseppe loves to read the menu out loud in his booming baritone voice, invariably rewarding your choice with a resonating *buonissimo*.

The local church and graveyard are situated not far from the crossroads, as is Montecalvo Versiggia Castle. More recent additions are the viewpoint (with benches and a mosaic of the county's coat of arms: a glass of *spumante*, sparkling wine), the small chapel for the Madonna of the Grape Harvest and the Corkscrew museum, world famous in Montecalvo and its neighbourhood. Crocetta is one of the 60 or so *frazioni* (hamlets) that form this agricultural county. All these *frazioni* names have historical origins (sometimes named after *one* specific house) and when listed in a row they make for colourful reading. Bagarello, Borgogna, Bosco, Ca' Bella, Ca' Galeazzi, Ca' Grande, Ca' Michele, Ca' Nuova, Ca' Rossini, Canerone, Capoluogo, Carichetta, Carolo, Casa Bassani, Casa Chiesa, Casa Galotti, Casa Ponte, Casa Sartori, Casa Tessitori, Casa Torregiani, Casa Zambello, Casaleggio, Casella, Casone, Castello, Castelrotto, Cerchiara, Colcio, Colombara, Colombato, Costa, Costiolone, Croce, Croce Bianca, Crocetta, Crocioni,

Fontanino Ninetta, Francia, Frenzo, Lanzone, Lardera, Marchisola, Michelazza, Moglialunga, Molino Nuovo, Mussolengo, Piane, Pianoni, Poggio, Poggiolo, Poggione, Pornenzo, Pratello, Remolato, Sasseo, Savoia, Schiavica, Spagna, Spinola, Stallarola, Tromba, Valazza, Valdonica, Versa, Versiggia. Our *frazione* is called Spagna, and its name derives from the nationality of a section of Napoleon's army that was stationed here for some time.

We had to wait about three-quarters of an hour for Roberta, who only spends part of her time in this small county; she was now travelling from a different county where she was working in the same post. She was delayed. It was worth the wait though because our *architetta* turned out to be very helpful and thorough. Files were dug out, maps unfolded and if something was unclear, she just shouted over her shoulder to one of her colleagues. Question: "It's OK for these guys just to start a B&B in their home, isn't it?" Answer: "Yes, of course, they don't need a permit for that." Or: "Have you got the maps for the land-use plan over there? Make a quick copy for these gentlemen here, will you?"

Luckily there didn't seem to be any insurmountable obstacles or ominous construction work planned for nuclear plants, motorways, waste incinerators or high-speed train tracks. We didn't even need to concern ourselves with earth quakes: according to the provincial regulations, these halted at the border of the county of Pavia. There was no official provision for them here.

Non si dà in prestito

The tall room of Pavia's university library was mainly occupied by people who sauntered along, seemingly aimlessly from one information desk to the next, in order to have a chat with a librarian at each one and then move on to the next. I saw one of these 'saunterers' near the exit with his coat on and a bag in his hand. Was he going to leave? Where would he go? But no, a bit later I saw him circling in the room again, right at the far end, without a coat or a bag. Where did he leave them? How did he move so quickly and invisibly to the other side? What did these 'saunterers' discuss, what were they actually doing here? But most of all: what was I doing here? Was I daydreaming?

Oh, yes, now I remember, I came to borrow a book. I climbed up the wide marble steps inside the university's dark corridor to arrive at the Central Library. The first-floor gallery of the university's atrium was deserted and the only thing in sight was an enormous and seemingly impregnable wooden door. I decided to have a go at pushing the massive door open and to my surprise, I could open it without any difficulty and go through. The door opened into a kind of waiting room, with rows of card catalogues and a desk with various personnel behind it. On the left, behind another door, I saw a large room.

A staff member behind the desk gave me a friendly nod and I walked towards them. I had to hand in all my stuff, my coat and my bag, and in return, I received a receipt. This was followed by a friendly, yet insistent stare. A lady asked me in a friendly manner what she could do for me. "Well, I would like to borrow a book," I answered hesitantly, ready to make a U-turn if it turned out that I was an intruder. But of course I was allowed to borrow, if I could show my university card and fill in a form. I was happy to oblige and my card was placed in a little wooden box. I saw that my card was the only one in that box. On the form, I specified, as well as I could the book for which I had made this long trek to the library. Title, author, year of publication, collocation,... I had no idea what they meant by 'collocation' so I decided to leave that box empty for now.

I was allowed through the door to the second room, the spacious, long room with the tall ceilings of the 'saunterers'. The side walls of the room were lined with cabinets filled with impressive-looking leather-bound books up to the ceiling, safely behind glass, out of the reach of unauthorized outsiders. An abandoned library ladder was propped up against the cabinets on both sides. Right down the centre there was a long corridor. At the far end of the room, I saw the door leading into the next room, but I couldn't see beyond it. On my left-hand side there was a wooden desk with three staff members behind it, who were all looking at me full of expectation. A customer! An outsider! Armed with my form I walked straight across to the desk. One of the members of staff examined my piece of paper with a serious expression. Something was missing. The collocation! They informed me that collocation referred to the book's call number: without that number it's impossible to locate the item. For the call number, I had to go back to the waiting room, to the card catalogues. But luckily I remembered just in time that I had scribbled down the call number on a piece of scrap paper. That was now in my coat pocket, which was hanging in the cloakroom, I realised with rising panic.

No, luckily, it was in my trouser pocket, and at least I still had my trousers on!

Now I had to fill in three forms which were more or less identical. Each one perforated, so that you could tear strips off. The three-headed librarian was observing my aching fingers as I kept on filling in the title, author, year, collocation and also my name and address and the name of the faculty. When I next looked up, I saw that something had changed: without me noticing, there had been a changing of the guard. These were not the same librarians any more. Where had the others gone? Had they joined the ranks of the 'saunterers'? One of the new ones took my forms and looked at them with some attention. He stamped them here and there and started to tear off the strips. Then he handed one of the strips to a younger member of staff, who disappeared with it behind an opaque glass wall. There must be a lift behind there because there was no other exit in the room. After some time, the same member of staff reappeared. With the book. After having checked the call number, I was allowed to take it, together with some of the forms.

I was directed to a desk on the other side of the path. There sat, slumped in his chair, another member of staff, wearing large glasses. He had a 'saunterer' near him, who hastily departed when he saw me approaching. From behind the thick lenses of his glasses, two watery, tired eyes looked up at me. I handed over my paperwork. He took a long time examining it and reached an uncompromising conclusion: *"Non si dà in prestito.* This book cannot be borrowed." Books which had my collocation seemed to be too delicate to be lent out to unauthorised outsiders. Such was my luck and now I went and returned the book to the first desk and left the library. The staff behind the desk didn't even blink. In the waiting room, downcast, I received my university card, coat and bag. Yet I skipped down the marble staircase, tore open the door and stood in the sunshine. I was free! I had to find a bar to celebrate!

Sono finiti i soldi!

Our lanky North-European bodies, stuck out far above the crowd of grey Italian heads. We were standing in the queue in the BancoPosta, the Italian Post office. Well, if you could call it a queue. There were several queues weaving in and out, tangled like strands of spaghetti. We didn't mind, we were not in a hurry because we had already spent days trying to open a *conto corrente*, a current account. We wanted an account with cards and especially a chequebook, because cheques are indispensable if you want to buy a house in Italy. When buying a house, one is required to make a number of advance payments which go direct from the buyer to the seller without any involvement from a solicitor. And these advances can be as high as 20% of the buying price. In other words, an amount of money that you would rather not carry in cash on your person. At least we don't, but the vulnerable elderly people standing in front of us seemed to have less of an issue with this because the cashiers handed them piles of fifty-euro notes counted into neat little bundles. They didn't seem to worry that they could be knocked over the head outside the post office by some rogue thief wanting to rob them of their pensions, despite the fact that the local newspaper was full of reports about mugging and burglaries.

Today was the first day of the month, the day when pensions were paid. This was the reason for the long queues in the Post office. Pavia is in fact a town of students *and* pensioners. Pensioners find relative peace and quiet here, together with good facilities. The students are attracted by the university and colleges. There are also a large number of secondary schools tucked away among Pavia's numerous hidden little squares, and on Fridays one is regularly taken aback by hordes of school children pushing their way forward towards the train station - not unlike the rats of Hamelin - leaving the city for a weekend at home.

We had forgotten it was pension day, but now it was too late. We really had to open this account after several futile attempts at trying to get our hands on one. Our odyssey had begun a couple of days earlier, in a local branch of BancoPosta. There it soon became clear, however, that one can only open an account at the head office. The head office occupies an enormous building in the city centre, but all counter service had been moved temporarily into a portakabin next door because of large-scale refurbishment. In order to apply for an account, you had to go to the little back room of the chief administrator. There sat Maria, who very kindly helped us to fill in all our forms. Within a couple of days we could come back to get our cards as well as our *assegni*, cheque book. We should be able to withdraw cash immediately, as soon as we had transferred the funds into our brand new account. And we had already done that.

Now finally, today was D-Day! Our attempt to get to our Madonna of the Post office ended in failure: a frightful looking officer was now guarding the entry to her Holiness. Apparently others had also discovered this little short cut and it had become over-exploited. There was no escape from the spaghetti queues today. We were patient: our salvation was near. As the queue in front of us slowly diminished and we came enticingly near the customer services window, suddenly panic broke out behind the desk. Someone shouted: "*Sono finiti i soldi*! We've run out of money!" This caused some murmur amongst the pensioner folk. The post office staff made it quite clear that there was really no more cash left. A couple of those entitled turned to leave gloomily, whilst others still tried to argue. But when the cash registers are empty, even an emperor couldn't withdraw his pension. After this incident the queue moved unexpectedly fast, and soon it was our turn. We received our cards and the cheque book, but we would need to return for our cash in a couple of days, when they had replenished their supply.

Gnocco fritto

"*Ahi,*" exclaimed Giorgio. He had burnt his fingers in the hot oil in the saucepan, whilst trying to fish out the small fried dough balls with bare hands. He had already admitted that he was not great at cooking and the menu he had chosen to cook for us tonight was selected on the basis of one important criterion: it was a guaranteed success. He was eager to impress us as this was the first time that we were visiting him at home.

The evening got off to a rather confused start. We thought that we had agreed to meet at his house, which stood in a new housing estate in a small village to the north of Pavia. We knew the address and the satnav brought us to the village, although its map hadn't been updated to contain the newly built area yet. So we were driving randomly in circles, looking at the street signs. We were in luck because it didn't take long before we spotted the street we needed. Now we only needed to locate number 10 and *saremmo a posto*, that's done. But that was not as easy as it sounds. The whole street had only number 10s: 10A, 10B, 10C, etc... And our host didn't provide us with a letter. We decided to ring him up, strange as it might seem, standing so near his front door.

"*Pronto,*" said Giorgio his voice sounding troubled. This sounded like the start of a chaotic conversation. Giorgio said that he was nearly on his way. "*Solo cinque minuti e poi ...*" On his way? We had agreed to meet at his place? Or hadn't we? Was he about to leave to go to our house? Or was he already nearly there? I felt the blood rising to my head. I tried to explain to Giorgio that I was really convinced that we had agreed to meet at his house.

Diplomacy was strongly advisable, because the well-known Dutch directness wasn't always appreciated, especially by such a sensitive Italian gentleman as Giorgio was. But, well, to find a tactful way of saying things, in a foreign language, under pressure, in a confused situation, that was a tall order. He didn't get me at all. After a couple of confused messages back and forth, I called out in panic that I hadn't the faintest idea, but we would clear everything up once we are at his house.

"*Ah, ma siete già qui*? Are you already here then?" asked Giorgio with astonishment. Ah, he finally got the message. "*Ma come è possibile*? But how is that possible?" "You know, we are just sitting in *la macchina*, the car, of course," I said, taken aback by his bewilderment. "But how were you able to find it, this street hasn't been recorded on any available maps yet!" he said dumbfounded. "You are amazing. Dutch people are so resourceful and rely on luck to find non-existent addresses with great success!" He had meant to pick us up in Pavia to prevent us from hopelessly getting lost, but the two crazy Dutch guys had already turned up at his doorstep. Unbelievable.

A bit later we stood in his apartment having a good laugh about all the confusion. Giorgio was still slightly stunned and amazed by how enterprising 'northerners' seemed to be. "You have already found out everything about the neighbourhood, been everywhere and know the place better than the locals!" he said. We nodded modestly, but were pleased to agree. We went to the kitchen, where we could carry on chatting whilst he would do his magic with his pots and pans. But doing more than one thing at a time is not easy, even if you have already prepared everything for the *gnocco fritto*.

Gnocco fritto is a dish from the neighbouring county of Piacenza, but it's also popular in the county of Pavia. They are deep-fried dough balls which are topped with the most delicious cold meats from the region, like for example, *pancetta* (streaky bacon), *coppa* ham (a cured meat, made from the nape and shoulder muscle of the pig) and *prosciutto crudo* (dry-cured raw ham). It is a simple but unimaginably delicious dish, the warm and airy dough (the pillows are hollow inside), *al dente*, topped with the cold meats. The miracle of Italian cuisine is - besides the glorious aromas - the creation of a tremendous sensory experience through the right combination of lukewarm, cool, sour, oily, hard and soft textures. Mmm, the *gnocchi* were a culinary festivity inside the mouth. Just be careful with the hot oil, Giorgio, *ahi*!

We were having a good chat and in the meantime Giorgio managed to prepare a stunningly delicious *risotto*, which he served with a surprising twist: he lined a large ladle with a slice of *prosciutto* and deposited a little pile of *risotto* on the top. He turned the contents carefully upside down onto a plate, which resulted in a little pile of *risotto*, beautifully wrapped in a slice of *prosciutto*. Success was indeed guaranteed!

After dinner, Giorgio's brother Franco dropped by to add to the entertainment. "*Tutto bene?*" he greeted us in his booming voice. "Yes, yes, *tutto* is *bene*!" we mumbled. The deep and meaningful conversations that we had been conducting with Giorgio, suddenly became lightened up by the often random offerings from Franco. "What are you supposed to do if you are at home and you hear a burglar?" he asked. "You must of course switch the TV on and turn it up to full volume; they won't be expecting that and will be so frightened that they will immediately beat a hasty retreat." Franco was beaming from ear to ear, satisfied with himself for contributing this pearl of wisdom, and nestled further into his armchair. "What should you do if there are people loitering near your apartment?" Franco didn't even wait for our answer; he was so smug about his newly acquired knowledge: "Light some fireworks, and catapult them at them!" Giorgio listened to his older brother's announcements with concern. After Franco had left, he confided his opinion to us: "Oh, Franco with his 60 years. He may be close to retirement, but *è sempre un bambino*, he is still a child at heart."

Proposta d'acquisto

All the details had been checked, the town hall's land-use plan had been examined, our *geometra* Buttini couldn't find anything wrong with the house, we had an Italian bank account with money in it and we were in possession of a cheque-book. In short, we were ready. Were we really going to go ahead with this? Buying a house in Italy?

These are the things that may befall you when you toy with the idea of emigration and starting a B&B abroad. You start to look around 'with no strings attached', you imagine all the criteria that your dream house should fulfil, come up with a long list of requirements and then, with some luck, it might just turn out that a house like that really exists. What do you do then? You have been cornered. It's now or never. If you don't dare to make the leap, if you now decide not to buy that house, you may just as well give up the search, because you are obviously not really up for it. You have been unexpectedly confronted with your dreams, and now you must face what they mean in reality. Yes or no, that is the pressing and suddenly unavoidable question. To buy or not to buy? To emigrate or not to emigrate?

There is only one last chance to put off making the final decision: you make an offer on the house and you wait and see how the owner reacts. You can be assured of a counter bid, and if you refuse that, then you still have time to change your mind. This sort of Russian roulette is a dangerous game, because if the owner unexpectedly accepts your first offer then suddenly you find yourself committed. In principle. Funnily enough, this pretend postponement of a definitive commitment made it easier for us to arrive at a decision. We decided to make an offer and see how it would all pan out. In the end, this was a leap of faith no matter how well prepared we were.

We e-mailed our bid together with a list of reasons why it was lower than the asking price (i.e. all the house's shortcomings which we would need to fix, and their associated costs) to our master estate agent Olita. There was, however, no response, not even a confirmation of receipt. Biting our nails, we sent the e-mail again, this time to the manager of the estate agency. We received an automatic response, saying that our e-mail had been opened, but still no proper answer. The next day, Olita rang us and rattled at us in fluent Italian. What was he actually saying? Had they accepted our offer? Were the owners even aware that we had made an offer and how much it was? We didn't quite get the message, but there definitely didn't seem to be any *problemi*.

After a while we realised that Olita was inviting us to the estate agency so that we might complete a formal offer: a *proposta d'acquisto*. In Italy, house buying involves a lot of paperwork. We were happy to sign the proposal, but we were wondering if we could first receive a draft by e-mail so that we could read what we were committing ourselves to. Olita couldn't imagine why we would need that, because *non ci sono problemi*! Luckily, one of his colleagues was more understanding and he sent us a *bozza*, a draft, that we showed to our own expert: Buttini. According to the *bozza*, if we signed the *proposta*, we had to pay an advance on the *caparra* (the 10 percent of the offer to be paid in advance when signing the purchase agreement). So it was an advance on the advance. How much of the 10 percent did we have to pay up front now? We had no idea. We were reluctant to call Olita again: he probably didn't know himself and would just pluck an amount out of the air. In that case we could just as well come up with a sum ourselves. We chose two thousand Euros, because that was about 10 percent of the 10 percent *caparra*. We thought that sounded a pretty reasonable first payment. But when we arrived at the estate agency and showed Olita the cheque, he didn't seem happy: "*Impari,*" he kept on repeating. "*Impari!*" What did he mean? We looked at each other, puzzled: do you get it? No. Later we came to understand that he didn't like our sum because it couldn't be derived easily from the rest of the outstanding payments (the rest of the advance and the rest of the purchase price). To us it seemed like such a nice round number. But Olita was a qualified accountant, registered with the Italian Chamber of Commerce! He had shown us his accreditation earlier with great pride: "See how clever I was when I did that?" It definitely looks impressive on paper, I will give him that, but proficient in calculations? Really?

The *proposta* also mentioned the commission received by the estate agent for their role as intermediaries. A thorny issue, because we are talking about a couple of percent points of the purchase price, in other words: thousands of euros. Our experiences with Olita didn't make us feel like rewarding him with large amounts of money. If we had had our way, we would have paid him nothing, but in Italy that's impossible. The estate agent, representing the seller, is by definition also the broker of the buyer: an impossible situation, but this is how Italian legislation works. The broker gains financially from both the seller and the buyer! At least 3 percent (per client!) seems to be the norm. Luckily, our good old Buttini produced an article written by an agent and broker interest group, which stated that the legislation doesn't prescribe a set minimum percentage. The commission is negotiable!

Armed with the article, we marched to the estate agency. It was immediately clear that the issue of 'commission' was top of Olita's priority list. Naturally, 3 percent was the desirable percentage, because that was what the legislation required, maintained our agent. I got Buttini's article out and pointed at the paragraph which we had highlighted in bright yellow. It contained the written confirmation by the broker and agent trade union that there is no legally established minimum commission in existence. We offered 2 percent and not a penny more. Olita's face clouded over and even his colleague, who shared the office with him, left the room looking annoyed. What were we thinking, didn't we realise how much money was spent on advertising? Every day, from early morning to late at night, Olita would drive up and down to help people find houses. Family life? Oh no, an estate agent had no time for that! His poor neglected children! Italian melodrama... And by the way: 3 percent was what it stated in the legislation, Olita repeated. I pointed again at the paragraph in the article. Highlighted in bright yellow.

In the end we won the plea, mainly because the article was hard to ignore: we paid 2 percent instead of 3. With the provision that at Christmas we would take Olita a present in his office to show how pleased we were with his services. We thought this was a strange request, but he seemed to mean it entirely seriously. Maybe his abilities were called into question in the office too and he needed to prop up his professional status through positive feedback from his clients? Whatever the reason, we had no difficulty accepting the deal: two bottles of wine for 1 percent commission, it's a no-brainer. Yet we didn't feel like celebrating because it still felt like giving a whole lot of money to someone who hadn't really earned it. Later we heard that the sellers had paid the 3 percent, no matter how they felt about Olita's 'services'.

After all the haggling, we signed the *proposta* and with that, sealed our offer. How would the seller react? The ink was not dry on the *proposta* yet and the telephone on Olita's desk was already ringing. Completely by coincidence, the seller was getting in touch. Olita gave them a full account in incomprehensible Italian. Unfortunately the owner didn't immediately make a counter bid. But he would come into Olita's office to talk about it that evening. We sat at home on the edge of our seats waiting for the news, but a bottle of *spumante* was already in the fridge just in case...

I pattini d'argento

Bart, Klaas, Piet, Hans, Jan, ... a series of typically Dutch names were lined up for scrutiny. "How *short* they are, *monosillabi*!" exclaimed the company around the table. I barely met half of our party (all of them ladies) during my four month stay at Pavia University. And the rest I met maybe three times or so. I was attending our research group's New Year's Eve dinner, again at the Osteria alle Carceri, in spite of my friend Giorgio's strict instructions never to eat there. As a *buongustaio*, gourmet, he knew all the best places to eat out in Pavia and its neighbourhood and this *osteria* was definitely not on his list. My earlier experiences in the company of *la professoressa* Chiara and *la* Nagel confirmed Giorgio's judgement, but whatever, the research group must have thought that this was the only restaurant in Pavia or maybe they had done a deal with them, I don't know.

I sat at the head of the table and most of the conversation went over my head because my listening skills still had plenty of room for improvement. The most important subject of conversation, as far as I could make out, was the pregnancy of Gabriela, a PhD student in the faculty, and what she would call the baby. It was going to be a girl, she knew that already, and she was going to call her Lucia. From there, the discussion turned to good children's names and to Dutch names, which to the surprise of those present are very short. Most Italian Christian names have at least two syllables.

Suddenly, one of my colleagues accosted me directly in an entirely non-Italian manner and asked me what I thought of the faculty. She asked this with a friendly expression on her face. "Erm,..." I was gasping for air and trying to come up with an appropriately vague answer. "*Buono, buono,*" was all I could squeeze out in my embarrassment. The truth was that I considered the faculty as a sleepy, uninspiring, dull group of people who had given up trying to conduct original research years ago. Only the most necessary activities were still carried out, albeit reluctantly: delivering one teaching module per year, supervising a couple of PhD students, attending a conference now and then. But I didn't want to share my honest opinion so bluntly over the dinner table. "*Potresti essere sincero, però,*" responded my colleague in an even less Italian manner to my feeble reply. Her utterance was accompanied with an affronted expression on her face. "You could at least be honest." She obviously shared my opinion, the opinion I didn't dare to express. At this point, I broke out in a sweat, but luckily the waitress rescued me, when she came to take our orders, after which the conversation took a different turn. Phew!

Soon I was presented with a greyish-white piece of fish on my snow-white designer plate. This time I hadn't asked advice from the Carceri experts instead, I had made my own choice based on a lot of deliberation. *Orata*, bream, is available in all supermarkets for a pittance, and it's always tasty even if you don't do much to it. It's enough to season it with a bit of salt and pepper. A safe choice, one would hope! But I hadn't taken the dungeon cook's incompetence into account because believe it or not, the *orata* tasted of nothing and was barely distinguishable from the white risotto I had last time. I ate with *'denti lunghi'* and was hoping and praying that this time they wouldn't ask me if I enjoyed my meal ("*buono, buono*").

And then suddenly, I don't know how this came up, but people were talking about a Dutch children's story, 'The

Silver Skates'. A story about a brother and sister who join a race on wooden skates (poverty trumps everything) and - what a miracle - they win the first prize: the silver skates. *This* story sums up the Netherlands, according to the group. I was not familiar with this story and told them tentatively that it was not well-known in the Netherlands, so maybe it's just made up. *What*? On receiving this piece of information, everyone became disgruntled; surprised and disappointed in equal measure, like a child who had been told that Santa doesn't really exist. They were deprived of an illusion. Everyone in Italy knew the story, they grew up with it and were convinced of its genuineness. The Netherlands was the location of *'I pattini d'argento'!*

It was only later that I realised that the story of the silver skates was the same as the story of Hans Brinker! To get the facts straight: in the original story, written in the 19th century by the American Mary Mapes Dodge, Hans Brinker is not the little boy who plugs the dam with his finger, but he is the brother of Gretel (yes, as in *the* Hans and Gretel, how did they come up with this) with whom he wins the silver skates. In the book, there is an anecdote about the finger in the dam but it's an heroic deed performed by another boy who is not named. This small section in the book later underwent a 'spin-off' process and the anonymous boy received the name of the main character, Hans. It seems my university placement had taught me something after all!

Il compromesso

Olita was sitting in his Sunday best behind his desk in the estate agency, dressed up neatly in a suit and tie, with his hair cut and combed. He was more nervous than us, even though we were the ones who were going to spend a couple of hundred grand to buy a house. Maybe he was worried that we would carry on haggling to give him an even lower commission? If that was the case, he'd probably better not go home to face his wife that night! The sellers, Mr and Mrs Colombo, and their daughter, together with an authorised representative of their former son-in-law (the daughter was divorced, but her ex still owned half of the property) sat to the left of the desk. The two of us, the future owners, sat on the other side. Colombo's dog and our Saar had already had a scuffle and we had to keep them well out of each other's sight. This didn't bode well.

On the other hand, what could go wrong? The Colombo family had accepted our formal offer in the *proposta d'acquisto*, they had immediately cashed our cheque with the advance on the *caparra* (*impari* or not) and now we were ready to sign the *compromesso*, preliminary sales contract, one of the many stages of house buying as the Italian method requires it. The contract must be accompanied by a cheque for the *caparra* (why did these terms always remind us of the Sicilian *mafia*?), which is the deposit adding up to 10 percent of the purchase price less the advance on the deposit. With painful effort, our qualified accountant had calculated the outstanding amount of *caparra* we still owed after we had signed the *proposta,* and we had a cheque ready with the correct sum.

After Olita had conducted some further necessary formalities, we were all ready to sign, easy peasy. In spite of the animosity between our dogs, the sellers and we seemed to hit it off, which was possibly due to the presence of our common adversary and 'representative' Olita. He insisted on reading through the whole document word for word, even though neither parties found that at all necessary. Yet we could hardly accuse our agent of being a perfectionist, since he still managed to make mistakes when filling in a couple of basic information boxes on the forms. The sellers were suddenly called Colombio, which prompted Mr Colombo to mutter extensively under his breath. "Colombio, Colombio, who is that?" Even the Italian national insurance number, the *codice fiscale*, was incorrect, even though this number can be derived directly from someone's personal data. Luckily there was a secretary on standby (*pronto soccorso*) who was in charge of creating a new, corrected *compromesso*.

Whilst we were waiting for Olita's 'homework' to be corrected, my glance fell on his bag. It looked like a doctor's bag, maybe containing a stethoscope, a reflex hammer, and other similar instruments. Out of boredom (or maybe nerves?) and to break the silence I made a comment on the bag. Olita explained that it was his portable office. He liked to have everything at his fingertips. "In that case, we might find Mrs Olita hiding in there too", I observed merrily. Olita threw me an annihilating glance. Mrs Colombo, on the other hand, giggled and suggested having a drink together somewhere after we had finished, adding that it should be the 'organiser's' treat. Olita wouldn't have any of it.

Finally, we were allowed to sign the finished documents. After at least twelve signatures had been collected from those present, the temporary contract became a legal document. Anyone breaking this agreement would have to pay a fine of 10 percent of the total purchase price (a sum equal to the *caparra*) to the other party. Naturally, none of us would want to do that. After the 'ceremony' was over, Olita led both parties out separately, so that he could discreetly create an opportunity for the sellers to pay his commission (3 percent), followed by our turn to pay him our 2 percent, in the form of a thick pile of 50 euros. That *mafia* feeling had crept up on us again.

Non sta bene

Cinzia, the secretary of Pavia University's Faculty of Medieval Culture looked at me with a serious expression. *"Non sta bene,"* she answered in a grave tone to my question where *la professoressa* Chiara was. "She is not doing well." I lost count of the times I had come into the university only to find the research group area closed. The shared research space, on the first floor of the historic university building, was on a mezzanine with a view of the atrium and its fountain garden. It felt like being in a salon. The space was closed off by a wooden frame with glass panels through which it was possible to catch a glimpse of several desks and tables belonging to PhD students who were largely invisible (maybe non-existent?). The door that gave access to this *salotto* was always locked. I just pressed my nose against the window and came to the conclusion that the *salotto* was once again deserted. All the highbrow books on the shelves along the wall looked at me with longing. "Take me! Read me!" they seemed to call. But unfortunately that wasn't going to happen today either.

This time I decided to go and ask for help from faculty support, i.e. the secretaries. Numerous support staff were always present, ready to support somebody if they were suddenly in need of support. But now as I took in Cinzia's ominous expression, I thought I finally understood why my supervisor never seemed to be there: she was probably seriously ill (cancer?). Oh dear. *"Che cosa c'è?"* I asked trying to seem neutral. "What's the matter?" *"Ha preso un raffreddore,"* said Cinzia, remaining deadly serious; the professor has a cold. Oh no, I just hope she will recover! Italians and their obsession with health...

Dejectedly, I traipsed along to the *stanza dei dottorandi*, the study room, which was at the other side of the building: another small box room for PhD students, consisting of two desks, one of which belonged to a professor of Classical Philosophy who must have received his PhD ages ago. Presumably, that's the reason I never had the good fortune to make his acquaintance during my stay, even though I was earnestly requested to vacate the room on Wednesday mornings because that was when the professor came in to engage in his classical studies. Without thinking, I inserted the key into the lock of the PhD study room when I suddenly realised that this was the professor's special morning. Carefully, I pushed the door slightly open and peeped through the gap. Was there really a stuffy old classicist sitting behind that accursed desk, feeling violated by my presence? But I need not have worried, there was no professor sitting there (perhaps he had a cold too?).

Gaining right of entry to this study room had cost me a lot of trouble, and it required putting pressure on *la* Nagel who was my supervisor and who I once succeeded in cornering in the *salotto*. When I had visited the department, the summer before our arrival in Italy, I had been welcomed with open arms, once the confusion surrounding my gender was cleared up. I was always welcome, I was told, there were plenty of desks and PCs and I would receive my own key to the *salotto*. My spouse (m/f) was likewise welcome and I could also bring our dog if I so pleased. Great! When I approached *la* Nagel regarding my own key the following September, however, suddenly there seemed to be a problem. "But I need somewhere to study?" I exclaimed in a panic. I couldn't see myself studying in Giorgio's flat with my other half Nico bustling about and our dog Saar constantly begging for attention. "Yes, of course", sighed *la* Nagel. You should talk to *la professoressa*, maybe the *stanza dei dottorandi* is available, though the faculty would also need to ask the Dean's permission in writing. And that could take some time.

Beyond all expectations, permission was soon granted, and a couple of days later I unlocked the door of the *stanza* for the first time with some excitement. What a stuffy box! With only one tiny skylight inserted high up in the ceiling, it wasn't unlike a prison cell. Well, on the plus side, it was at least dry and reasonably warm. I would accept anything to avoid shivering on a bench in a park. Although the study room, just like the *salotto*, was nearly always empty, the room next door was always occupied. I could follow the telephone conversations that were conducted on the other side of the wall nearly word for word. I also listened to many a researcher's conversations with students about exams, assignments, marks and appointments. Conversations which always ended in a litany of *ciao ciao ciao*s interrupted by one last piece of information, then resuming the sing-song of *ciao ciao ciao*s. Italians and saying goodbye...

About every two hours I took a break to stretch my legs. A little wander along Pavia's cobblestoned streets to have a coffee in one of the cafés. Cafés which were always full and lively with people (PhD students? *Professoresse*?) who apparently had nothing better to do. Some of them were, already at this hour, sipping sparkling white wine. *La dolce vita!*

I was honoured to meet *la professoressa* Chiara a total of three times in the end: at the welcome lunch in the *Osteria alle Carceri,* at a departmental lunch in the *Osteria alle Carceri* and finally at a meeting lunch, where else but in the *Osteria alle Carceri*. We made one appointment via e-mail to discuss my research project, but this was cancelled shortly later by Cinzia, the departmental secretary (sick aunt, *raffreddore*, cancer?). From then on, I never clapped eyes on my *professoressa* again.

Le vertigini

One day we took a trip along the hills of the Oltrepò after we had first reassured ourselves that 'our' house was still standing and wasn't washed away by the abundant rainfall of the last couple of days. Luckily, the house was still standing firm in its place among the knee-high vegetation: an unsightly grey cube that had still not received that lick of violet paint. Olita, you have work to do! Otherwise, there will be *problemi.*

We decided to take a little look around our prospective neighbourhood and we drove to the south, into higher terrain. The roads became twistier and narrower and we were climbing gradually higher and higher with each bend in the road. My insides started to protest. Had I eaten something bad this morning? The gingerbread cake we brought with us from the Netherlands? Of course, it must have been past its sell by date! I thought gingerbread cake couldn't go off? How could this happen to me? I am always so careful to check everything that passes my lips for signs of mould or wriggling life forms. Or was it the strong *espresso* from the *moca*, our little Italian percolator? It produces powerful coffee that requires a hardy stomach. I started to feel sick. This was followed by a vision.

I was standing on the steps of the diving platform in the swimming pool. I was trying desperately to cling on to the smooth, cold metal rails. I was scared that my feet would slip off the slippery steps, which would first break my leg and would then send me falling backwards head over heels smashing my head on the tiles. I couldn't get a proper grip on the handrail, the steps were a lot steeper than I had expected, and I started to shake. But there was no way back, I had to carry on climbing. Once at the top, I heaved a sigh of relief. Finally, I was standing safely on a wide flat diving board fenced off by metal mesh on both sides. I decided never to do this again. But now I would need to move forward, beyond the fencing, all the way to the end of the wobbly edge, and jump into the depths. I looked down and realised that from the top it was a lot higher than it seemed from the bottom. It made me feel light-headed. How did I get myself into this mess? Behind me, a small group of macho boys were forming, staring at me impatiently. I had to jump, so I closed my eyes, held my breath and... took the plunge.

We had to stop at the side of the road for a break, so I could get some fresh air and let my stomach settle down. What exactly had we got ourselves into? Moving to an entirely different environment, with foreign people, foreign customs and a foreign language, which we hardly spoke or understood. Removed from our social network in the Netherlands, removed from our snug nest where we had lived for the best part of 20 years and where we felt at home and protected. Away from our friends and family, who could support us through difficult times and who could share the joys of the good times. Were we really leaving this all behind? And for what? Panic seized my body. Imagined all the things that could go wrong. Was there still a way back?

After the first wave of panic had died down, I returned to my senses. No, making a U-turn now was not an option. We had already taken the first steps and there was only one way forward: to proceed without wavering. But did we really dare to? Once back in our flat, my stomach and my mind regained their composure. We were going to create an amazing holiday destination, I realised, in beautiful countryside, in fantastic climate. We would have lots of interesting guests and we would make sure they would feel just as at home as we do. Stop hesitating, close your eyes and jump!

Il fannullone

Saar barked and I woke up with a jolt. I thought I had only been asleep briefly because I had just heard the clock in our living room chime 4 a.m. I thought it was strange, though, that I had already heard a bus going past. But last night I had set the alarm to 7:45 a.m. so what could go wrong? Half an hour later, and still awake, I was thinking: oh, I'd better check the time. I pressed the light switch on the alarm clock under the duvet so as not to disturb Nico, and what did I see? QUARTER PAST EIGHT! We had to be at the solicitor's within the hour! Why did the alarm not go off?

I nudged Nico awake, jumped out of bed, had a quick wash and gulped down a bowl of cornflakes. Nico quickly let Saar out; our faithful Saar who was trying to warn us with her gentle bark that it was time to get up! In the meantime, I wrote out the cheque to Colombo for the outstanding purchase price. An enormous amount of *non trasferibile*. According to Olita, it was OK to use the same kind of cheque as the one we used for the *compromesso*. Just to be sure, I had rung him to confirm this in December. I found it a bit strange that a sum of money this large, the price of a house, could just be transferred over using an ordinary cheque. But naturally, there were no *problemi*. Said our expert, who cost us twelve thousand euros.

Whilst getting dressed, I had a quick double check of the alarm clock. The alarm was set to 7:45, but... was that 7:45 p.m.? It should be a.m. of course... That stupid Anglo-Saxon system! Luckily there was no harm done as long as we got out of the door quickly. How were we going to get there? It would take too long to walk. Cycling was too risky, because of black ice: the melted snow had frozen to the surfaces overnight. We decided to go by car. Saar, our little hero, stayed at home, to wait for the return of his masters, by then hopefully the owners of a house in Italy.

We arrived at the solicitor's office at the same time as the Colombos and we waved at each other from a distance. The office was located in the courtyard of one of the old *condominis*, in Pavia's city centre, in a small side-street off Strada Nuova. Once inside, we waited till 9:15 a.m. when a little, fragile-looking old lady appeared, her back crooked from having bent over too many documents. Without looking directly at us, she gestured for us to come in. Olita was not present, despite earlier promises that he would accompany the sellers as 'a friend'. He could not appear in his role as an estate agent because then it would be obvious that we had paid him a commission which is taxable. We deliberately didn't pay that tax to reduce the already astronomical transfer fees. And of all people, it was the solicitor whose job it was to retrieve this tax. Later on, we were told that Olita had promised Colombo that he wouldn't be far away and would be available in case we needed him; that he would even be sitting in a bar across the road from the solicitor's office. Where, of course, he was not sitting, as we would discover later that morning.

Without further ado, our *notaio* began to read aloud the prepared documents, *nomi, cognomi, codici fiscali* (always that *codice fiscale*!). Interestingly, in the document, it said that we were fluent in Italian. After she had read this piece of information, she looked at us over the top of her reading glasses with questioning (sceptical?) eyes. We could be silent in all languages, but we nodded solemnly to confirm that our Italian flowed like a river. I waved a hand at Mrs Colombo, gesturing that it was *'così così*, so so' which made her giggle again. In the meantime, I spotted Mr Colombo staring at my cheque book that was visible through the plastic folder holding all our documents. Was he already licking his lips?

The solicitor took her time. We didn't understand everything despite being 'fluent' in Italian. Some corrections were made. *Frazione Spegna*? No, Spagna. That was a mistake. Then the Colombo's mortgage was examined: was it already paid off? Yes, of course, declared Colombo. But the original bank statement wasn't here to prove it. No, Olita had it. Who wasn't here. Or in the neighbourhood. Nowhere to be seen or to be reached. And he hadn't attached the statement to the solicitor's dossier either. Would everything come to a standstill? Luckily we had everything, even a copy of the bank statement, and that seemed to be enough. The solicitor launched herself into a long speech but I gradually lost the thread due to the formal language. Slowly it dawned on me that she was talking about payments. Colombo was again taking a sideways glance at my chequebook. Yes, yes, it's coming, you will be paid, don't worry, I was thinking. But that's when they dropped the bombshell. Colombo wanted a guaranteed cheque, not an ordinary, everyday one like what you use for small sums. I could sympathise with this wish, that was the reason I had rung Olita previously, who reassured me at the time. Why did I have to listen to him instead of asking someone more qualified?

And at that moment, when Colombo started on about the guaranteed cheque, in my mind's eye I started seeing assorted ways of inflicting pain and death by torture (in a slow and excruciating way). The passive object of these executions was of course not Colombo, because he was entirely right, but Olita. He wasn't there, the *fannullone*, the good-for-nothing layabout, and now we knew why: his incompetence took centre stage today and he preferred not to bear witness to that.

Luckily our solicitor was accommodating and let me walk across to the post office, or rather the portakabin, to have my cheque guaranteed. The others went to have a cup of coffee on the corner, where Olita, our family friend, never turned up. This gave Nico and the Colombo family an opportunity to exchange experiences about Olita. "*Non ha fatto niente!*" they kept on telling Nico. "He has done nothing!" He never delivered on his promises to refund their costs incurred on his behalf. And the Colombos had paid the full 3 percent agent's commission, *and* given him a Christmas present: a bottle of champagne! When they learnt that we had managed to haggle the commission down to 2 percent, the Colombos were furious.

At the post office it was pension day once again, but luckily the queues moved quickly and the money didn't run out. I saw the same administrator as last time, the one who couldn't be of any service at the time. I just hope this goes well, I thought with quiet apprehension. I only want to have a cheque for a mere 200 grand guaranteed... But it was easier than expected: she asked her colleague a couple of questions and sure enough, after some paperwork, checking our bank account and some print-outs, I was holding a guaranteed cheque in my hands! Efficiency itself: long live the BancoPosta! I hurried back to the solicitor's office, where we were finally able to sign the contract.

At home, just as we were raising our glasses for a toast, Olita rang. I had already mentally prepared myself and gave him a piece of my mind in fluent Italian. It wasn't for no reason that I had declared under oath - in the presence of a solicitor - that I was a fluent Italian speaker. Not that Olita understood why we were so angry with him: there were no *problemi* after all? I let him pretend, but I decided that I would never see him again.

And this is how, partly with the help of our watchful and mollycoddled dog, we became the owners of a big box of a house in Italy.

I Pedra

"Our village has had a bad year," said the Lions Club chairman at the Theatre Martinetti, in the little village of Garlasco. It was the start of the traditional 'Festival of Good Wishes', the *Festa degli Auguri*. Following the murder of the beautiful Chiara Poggi in 2007, which had prompted an invasion from the national press, the village had woken up one morning to the shocking double murder (suicide?) of an older man and a young woman. Next year will hopefully be a happier one. The evening took off with this positive spirit in mind, with a combination of the Ticinum Gospel Choir and a performance by the folk music group *I Pedra*. Nico had recently joined the gospel choir and he was fast becoming a respected member because of his clear tenor singing voice. The members of the Lion Club must have been looking forward to this little diversion, since all the two hundred and fifty seats in the theatre were occupied.

The *Teatro Martinetti* was a nice little theatre. Like many other provincial Italian theatres, it was designed in the style of Milan's *La Scala*, only scaled down to model village size. The bagpipers and the choir were performing for a good cause tonight, because this year the festival was dedicated to those suffering from a mental illness: it was a *serata benefica a favore dei disabili mentali*. The charitable donations were collected at the door, where it was left to your own discretion how much money you wanted to contribute; it was *ingresso ad offerta*.

In Italy it's not uncommon for a theatre performance to start late. Sometimes by a quarter of an hour or half an hour or even forty five minutes... no one seems to take any notice. This gives you the opportunity to take a good look at the slow stream of people arriving. In the Theatre Martinetti I observed a lot of greying heads and an overwhelming number of old ladies in fur coats. Real fur. In Italy fur is still displayed with pride and without embarrassment and there are no animal rights activists standing on street corners with their spray paint ready to humiliate the owner. Everyone had made an effort and was dressed up to the nines. *Fare bella figura*, the well-known Italian adage that you should always present yourself at your best to the outside world, was gloriously at work. No worn-out jeans, baggy jumpers or scruffy shoes, only clothes cut in the latest Italian fashion. The audience stood around in small groups, chatting without showing any signs of impatience about the late start of the show. People were wandering around, some even leaving the room. Would the performance still take place or had there been a disaster that was going to put an end to the proceedings? These were the questions going through the minds of the punctual Northerners, which we still were. No, it was nothing like that, there was nothing wrong at all, this is just Italy. These Italians and their ways!

In the end the performance finally began. *I Pedra*'s bagpipers, dressed in folk costumes, came out onto the stage, but before the tootling could begin, the band was introduced by a speaker. The speaker could have come straight out of a comedy sketch, and he certainly took his time in the spotlight. His mumble was hard to make out, but that wasn't the point. *I Pedra* commanded the audience's full attention, not just because of the clothes they were wearing but also because of the faces that went with them. Suffice it to say, they looked a bit special in more ways than one. I couldn't recollect having heard a rendition of 'Silent Night' in a louder or more ear-splitting tone. *I Pedra*'s version of Brahms' Lullaby for the Christ Child would give any child ADHD. And the special key in which 'Adeste fideles' was performed made me grind my teeth. In the box next to mine, someone fell off their plushly upholstered chair from laughing.

Luckily the repertoire that *I Pedra* performed that night was restricted to five songs. This took the best part of an hour because the speaker had to make some comments after each song. Yet, the stage was the focus of everyone's attention. Centre stage was occupied by the man playing the triangle, whose expression went from imperturbable to very imperturbable to completely imperturbable. On one side stood a father and son duo playing the recorder, both of them short and squat, their cheeks bulging with air as they blew. Both of them wore a Paddington-style hat pulled over their eyes. Some of these images are now ingrained in our minds forever. When I think of Garlasco, I don't think of the murdered teenager or the couple who committed suicide, but I think of *I Pedra*.

The gospel choir's subsequent performance was a welcome relief to those with sensitive ears and delicate constitutions. They started off carefully, with a quiet piece, led by a soloist. As they carried on, and brought in a bit of a swing, the traumatised audience slowly came back to life. The conductor's enthusiasm was infectious, so much so that he succeeded in getting everyone to join in with the *'Kum ba ya my Lord"* bit. Time and time again, a soloist would step to the front and lead the choir in some English classics. Nico performed 'O Tannenbaum' as part of a multi-lingual medley. The performance was such a success that the choir was called back for several encores. In the end, we didn't leave the building until around midnight, feeling cheered up and exhilarated and ready to have our *aperitivos*, in the truly silent and holy night.

Servizi pubblici

Now that we were the proud owners of the house on the hill (we still had to come up with a name for our forthcoming B&B), we regularly visited the site to assess the amount of work needed, with regard to clearing up, cleaning and repair work. This was always an enjoyable outing, because the flat in Via Moruzzi wasn't suitable for staying indoors for long periods of time. The flat turned out to be quite dark, even in sunny weather, because all the windows were facing north and were in the shade most of the time. Although our balcony was quite large, it also faced north, and we could only catch the sun in one of its corners in the very early spring. It would be a shame to waste beautiful sunny days by staying indoors, so these were the days that we made our journeys.

Receiving the keys to our new house wasn't as easy as we had first expected. Because our estate agent Olita hadn't attended the transfer meeting at the solicitor's (absent without prior notification, shall we say), we still hadn't received the keys to the house. The Colombo family gave the keys to Olita, so that he could carry out his estate agent duties to his own amazing standards. So we could get our hands on the keys we had to pay yet another visit to that *fannullone* (slacker). I had no intention of ever having anything to do with the man again, so poor Nico had to pay the price. Olita, of course, immediately started up about our grudge against him because he couldn't understand what he had done wrong. The seller, Mr Colombo, was extremely satisfied with him, he insisted. We knew better, having spoken to the Colombo family ourselves. Olita needed to go through various drawers and cabinets before he finally located the keys. Hopefully these were the right keys, we were thinking, but luckily it wasn't necessary to pay another visit to the *fannullone* because this time he had actually got something right. He was worth his weight in gold!

Once we had driven past the Ponte della Becca, the landscape seemed to open up and become lighter. The first hills appeared over the horizon and vineyards were dotted along both sides of the road. On these mornings, the radio often played '*Back to Black*' by Amy Winehouse (an apt surname...) and we soon began to associate her music with our days out in the Oltrepò. To us, everything was still new, and we were wide-eyed with wonder. Sometimes, we spotted sumptuous *ville* at the end of long avenues, probably the family homes of vineyard owners who had grown grapes there for centuries. Cigognola castle's dovetail-jointed towers were looking at us invitingly from the top of the hill. We drove through the valley of Scuropasso, a *torrente,* a creek, that swells up considerably after the heavy spring showers, but for the rest of the year remains mostly dry.

By now we had discovered two routes to get to the house, two little roads winding their way up the hill, and alternated between them. Today we were going to see if we could get all the utilities to work: the gas, the water, the electricity and the telephone. Just before the official handover at the solicitor's, we had made a last check with *signor* Colombo. Was everything still in the same working order as when we signed the *compromesso*? Did Colombo behave as a *buon padre di famiglia*, a responsible family man, as it's so beautifully put in the sales agreement? Yes, he did. Until now we hadn't heard him speak much, but he seemed to be a friendly old fellow, who was willing to take his time to explain everything to us. Without signs of hurry or impatience.

Even the *rustico*, the little two-storey brick building behind the house, had gas and water in it and in order to get electricity installed, it was enough to make a quick phone call to ENEL, the energy provider, Colombo assured us. The spacious cellar under the house, the *cantina*, was brimming with water connections because originally they had planned to make a laundry service in there. There were two thermostats so that you could control the central heating in both flats separately. The importance of this *riscaldamento autonomo* was by this time quite clear to us. Our stay in Via Moruzzi had left us with 'sauna-syndrome': in the depths of winter we sat with the windows wide open because the inside temperature was near enough 30 degrees Celsius. The heating was centrally regulated for the whole *condominio* and it was permanently on 'high setting'. This is why properties-for-rent adverts put so much emphasis on the availability of *riscaldamento autonomo*!

The cellar and the *rustico* both seemed to serve as dumping grounds for unwanted furniture, building materials and other junk: a loose toilet bowl, a bidet, a marble sink, a whole bed, cabinets, cabinets, cabinets, a big pile of roof tiles, enough for a whole new roof: in other words, a total mess. Whether or not we were supposed to be grateful for all this, we couldn't yet tell. In a hole under the drive there was a pump, said Colombo, which pumped the waste water from the kitchen on the first floor to the waste pipe which was in an elevated position on the other side of the house. And the waste pipe was connected to the main sewage pipe which ran along under our street in front of the house. All the bathrooms were connected to a septic tank, which was again connected to the sewer system. *Nessun problema*! The electricity had recently been rewired, and complied with current regulations: *tutto a norma*. This is what Colombo had said at that time, right before the hand over.

After a pleasant half an hour's drive, we reached our big grey colossus and parked our Fiat Punto on the concrete drive. Every time we opened our big, burglar-proof door, we noticed how cool it was indoors. We secretly toyed with the idea of spending a night in our new property, but the air was filled with the unpleasant sewage smell of toilets and sinks which had stood dry for too long. Somehow we had to manage to get the water connected. A helpful gentleman in Montecalvo town hall had given us the numbers of the water, gas and electricity providers. We had to take a deep breath before we tackled the job of ringing a number of different call centres. We were worried that besides the language barrier, the infamous unfriendliness of Italian energy providers would also complicate things. It was going to be a high-intensity Italian course, which would put our blood pressure up. We had prepared a couple of standard sentences in advance, so we could explain to the ladies in the call centre exactly what it was that we wanted. At least, that was the idea. The *callgirls* did seem to listen at first, but not for long, soon they would launch into fast-paced monologues and start connecting us to random colleagues who did the same: holding their breath for 20 seconds, followed by a fast machine-gun salvo of words. Sometimes, in the midst of this verbal torrent, we made out a recognisable question or two, like for example the one about the post code of Montecalvo. We didn't have it ready, and by the time we looked it up, the call centre lady in question had already hung up.

In another attempt to navigate through the cryptic bureaucracy of the energy providers, I tried speaking in English. On the other end of the line you could hear the panic gripping them. Do you speak English, I politely asked first in Italian. "*Uh lietl…,*" was the not very encouraging answer. So it had to be done in Italian. In the end, it started to go a bit better and the call centre ladies took our attempts more seriously. After every phone call we found out what information we had to get ready for the next one, slowly making progress. After we had provided all the necessary information, including the *codice fiscale* (which by now I could spell in Italian: Bari-Torino-Salerno-Napoli-…), I was told that in five days we would have *luce*, electricity. Now for the gas and water.

When talking to the gas provider (there was a new number for the ENEL too) an additional problem arose: the *contatore*, gas metre seemed to have been removed from the house. Was that now completely *a norma, buon padre* Colombo? The *tubi,* pipes were installed, but a new metre still needed to be connected. After some more discussion and a long wait on the line whilst cheerful Italian *canzoni* were blaring in my ear, we agreed that they would come out on Thursday afternoon. Requesting the connection of the *acquedotto*, water main, was surprisingly easy. The operators understood us immediately, they knew where Montecalvo was and were even familiar with *frazione* Spagna. Unfortunately they informed us that the house had no water *contatore*, either. Our understanding of what belongs under the scope of *a norma* was clearly very different from that of *padre* Colombo. But first things first: we were required to pay before we would get any water. "Come over tomorrow morning to Stradella. Before 10 o'clock. Then we can sort the contract and the payment out," shouted the voice at the other end of the line resolutely. Then the water metre can also be connected on Thursday morning. That sounded like another nice trip out to the Oltrepò! I could already hear wine-girl Amy singing: *"We only said goodbye with words ..."*

Di fiducia

"*L'ingegnere Cassani,*" said Franco, without a moment's hesitation. We had asked him if he, as a *geometra* of the county of Pavia, could recommend us a *di fiducia* architect or architectural engineer whose advice we could trust. His answer was firm and satisfactorily convincing too. It would have to be him. We wanted an expert to design the structural adjustments that needed to be made to our newly bought house, because in this respect we were complete *dilettanti,* we knew next to nothing about house extensions. The same architect or architectural engineer could also supervise the execution and quality of the building project and manage the contractors. We were not very keen on project management ourselves.

But how do you find someone like that? By picking a random name from the *Pagine Gialle*, Yellow Pages? That didn't seem a safe option to us. So we decided to do it the Italian way: use your network and make sure you find someone *di fiducia*. This typical Italian system works as follows: you ask someone you trust whether they know someone they trust. If they do, you are sorted, because an Italian would rather die a thousand times over than betray your confidence. If they don't know anyone suitable, then they will ask around their trusted network and if a name comes up from there, you will be fine. Everything in Italy works the same way: you have a GP *di fiducia*, a plumber *di fiducia*, a mechanic *di fiducia*, etc... In fact, the reality is that an Italian will trust no one, unless they are *di fiducia*. The world is one big jungle full of swindlers and crooks. Or at least Italy is.

Luckily we had Franco, who would certainly not want to betray our trust and who would be able to recommend us a good architect or engineer from his professional circle. Cassani it was. We made an introductory appointment and went to find his office in the centre of Pavia. Our engineer was a modestly attired gentleman with gold-rimmed spectacles, wavy hair and a well-groomed beard. His powerful voice suggested that he was someone possessed of a great deal of competence. His assistant was a quiet and rather surly-looking balding man with a grey beard. He sat to one side, behind Cassani (who occupied the central position behind his enormous desk) and muttered unintelligibly once in a while. We explained as best we could in our broken Italian what we had come for: a couple of small modifications to the house (some extra windows), a terrace, an indoor staircase to the cellar, which would need a new entrance door. And we wanted the roof fixed because we saw that some rooftiles were missing. In some places the sky was showing through the roof. The building was in good condition, our *geometra* Buttini had established that during the *perizia*, and we wanted to keep it that way.

We decided that the roof was priority, because the winters are often rainy and very windy. Cassani immediately recommended a *di fiducia* roofer, who owed him a favour too and would be prepared to start the job soon. Cassani phoned him and picked a date for the *sopralluogo*, a visit to assess the situation on site. He would then take a look around and see what was expected of him. Was the assistant going to accompany him? Yes, he was coming, although he didn't seem keen on the prospect.

Il water

In Italy, just like anywhere else, you need to visit the public facilities (at bars, restaurants, theatres, universities, etc) now and then. This can become quite an adventure for the toilet trainees just starting out, although even advanced toilet-goers experience regular set backs.

To start off with, you need to locate the toilet. If the *segnalazione*, signs, are not satisfactory (and that is often the case), you are faced with the task of asking a member of staff or others present to direct you to the shortest way to the facilities. "Could you tell me where the toilets are?" How do you say this in Italian without making a blunder? The notion 'WC' won't cut the mustard here, even if you knew how to pronounce this in Italian (*"doppio vee tshee"*). With a cry of distress *"Toilette?"* you will have more luck, even though uttering this single word doesn't make a very good impression. Italians call the toilet *il bagno*, an expression that's often misused by Dutch people who are trying to find the bathroom. Because Italy is essentially the country of gesture, theatrical hand wringing will suffice too.

Once you arrive at the toilet, you have an existential choice to make: man or woman. Even if you you are in no doubt about this (you can always check your passport), you may still encounter another obstacle: commonly, there is no clear picture of a little man or a little woman on the respective doors. With some luck there might be some writing: *'Signore'* and *'Signori'* on the doors. But even this can often give rise to panic. Which is the one for 'Ladies' and which one is for 'Gentlemen'? They both seem masculine, don't they? For heaven's sake, why can't they write *'Donne'* on the door of the women's one? That would at least be recognisable to the uninformed foreigner. You could hang around nonchalantly until you see someone disappearing into or reappearing from one of the doors, and use that information as an indicator for which one is which, but what to do if your need is urgent and the toilet traffic 'sparse'? This is why you need to know (and don't forget this!) that the feminine plural of most common Italian nouns ends in 'e' and the masculine plural usually ends in 'i'. There are exceptions, but with regard to toilet facilities. There the rules are strictly observed. Even in Italy.

Finally you can drop your pants. Whoa, hang on, not so fast! Experience teaches us that you never quite know what you're going to find behind the toilet door. First of all the question is how complete is your privacy? Can you actually lock the door? No, often the original lock is out of order but if you are lucky the door can be locked by a DIY mechanism (hook, string), improvised by someone in need. In most cases, there is no such luck and there is no key, the whole hanging- or hooking- system is gone or maybe there was never a lock in the first place. Anything is possible. Even in brand new toilets in shiny new buildings you can find yourself up the proverbial creek without a paddle: a toilet door fitted with a traditional lock... but without the key. If the door can't be locked but the cubicle is so small that the toilet is close to the door, it's still 'not the end of the world'. As you are sitting you can hold the door shut with a hand or a foot. In case, against all hope, someone happens to walk in, you keep hold of the door and shout out anxiously: "*Occupato!* Occupied!" (something to remember!)

Sitting? That's the next question: can you actually sit down? Although most Italian toilets are of the 'bowl' variety, you can still come across the occasional *bagno alla turca*. This may sound like an exotic oriental piece of music, but in reality it is the infamous 'hole in the ground': the squat toilet. If the door can't be locked and you discover one of these Turkish squats behind it, you have no hope left. At this point there are a number of possible scenarios, one more humiliating (for the victim) and funnier (for the spectator) than the other. I will leave it to the aspirant toilet-goer to use his/her imagination to fill in the details. If the door can be locked, there are still plenty of potential stumbling blocks.

Using the Turkish beast can lead to a lot of fuss involving trousers, underpants, vests and shirts, as well as a balancing act to stop you disappearing through the hole (although we are not that skinny any more). Successful completion of 'business' depends on a laborious co-operation between the producer hole and the recipient hole. Cramp in your thighs, calves and tendons is a certain consequence; wet socks, trousers and shoes are a very real possibility resulting from these efforts. Or as one Internet user guide has put it:

>Squat toilet
>*Entertaining - but slowly disappearing - phenomenon in Mediterranean countries.*
>*The use of the 'gabinetto alla turca' is very hygienic as long as one knows how to aim. For people without experience, the adventure often ends in wet socks.*
>*Tips: try, instead of only bending your knees slightly, to lower yourself completely into a squatting position, and keep your head facing the door.*

The last piece of advice contributes the necessary hint of humour and relaxation in this difficult predicament: if the door unexpectedly swings open, you can laugh the laugh of the innocents, because your head is pointing towards the door. But it can get worse (there is a type of squat toilet which stands on a small platform which makes it impossible to squat above it without completely removing your trousers. Imagine the door flying open whilst you are occupying that one...).

Thank goodness, there's a toilet bowl, you think to yourself, if the dreaded squat toilet doesn't materialise. But what kind of a bowl? Is there a toilet seat? *Il water,* as the toilet bowl is called in Italian (derived from water closet) is rarely equipped with a seat (an *asse*). If there is a toilet seat, it will be standing sadly in a corner (as a punishment), or under lock and key on the wall (no touching), or it's broken (with chunks missing), or it's loose, or it's too wet and disgusting to sit on. The last option is often a result of the fact that most toilet seats can't be put up: they promptly fall back down into place because the cistern is in the way. We must conclude that the Italian doesn't get the purpose of the toilet seat: he regards the toilet bowl as a noteworthy variation on the squat toilet, one which is raised but which is not meant to be sat on. Who came up with something so stupid, you can hear him think.

With some luck, we might be able to improvise something near to a sitting position and perform the big deed in peace. It will have to happen in the dark, because the energy saving system is usually activated after 30 seconds, the on/off switch is out of reach, or the motion detection sensor is broken. If you have spent your penny by then, then the worst is over. There is often no toilet paper, but of course you've already taken care of that, because you've learnt from past experiences (always check *first* if there's any paper!). The Italian doesn't use paper, that's what the bidet is for. That crazy little bath that Dutch people use to wash their feet in. It's a shame that in precisely those places where they are the most needed, i.e. in public toilets, you will hardly ever see a bidet. The rest I would rather leave to your imagination.

Flushing takes some imagination too: push buttons on the cistern are either broken or have been replaced with buttons on the walls or some sort of foot pedal. And sometimes the automatic cleaning system gets activated and the toilet seat starts turning underneath you whilst you are still busy, but only a spoilsport would let that bother them. Washing your hands, as you always promised your mum you would do, is not always possible. Soap is rare, although water is usually available. Sometimes you can operate the tap very hygienically with foot pedals, but don't count on there being any paper towels to dry your hands with. Or maybe there are, because it's possible that you have washed your hands at the wrong sink. Interestingly enough, Italian public toilets often have two different rooms with sinks and only when you are leaving do you realise that the second room did have soap and paper or an *electric hand-dryer*. Then you also suddenly find out that the door of this toilet *can* be locked. Going to the toilet in Italy remains a lonely and stressful exploit!

I copritetti

Autista, antennista, farmacista, barista, giornalista, ... It's hard to think of a profession that doesn't end in *-sta* in Italian. *Autista*, for example, does not usually refer to a person with autism, but instead to a professional driver, someone we might call a 'chauffeur' at home. The sign, *Non parlare con l'autista*, often displayed on public transport, has a very different meaning from what it looks like at first glance... An *antennista* is a person who earns his living by putting up aerials on people's roofs. They really exist! Furthermore, an Inter Milan supporter is an *Interista*, an accountant is a *commercialista*, a dentist is a *dentista* and a tyre salesman is a *gommista*. These are all masculine words even though they end in the feminine marker *-a*. As Italian 'novices' we made a game out of coming up with new variations. You pick a profession, add *-sta* at the end and you are done: a legislator is a *leggista*, a pizza baker is a *pizzista*, etc... All wrong, because when you think that you've finally got it, it turns out to be different. Like so many things in Italy.

We had an appointment with the roofer who was rustled up by our architect, Cassani and we expected him to be a *tettista*, derived from *tetto* (roof) + *sta*. Wrong! A roofer is called a *copritetto*, literally a coverer of roofs. Whatever he is called, he seemed to be earning a good living because after we had been driving behind him for a while, we realised that he was driving a Jaguar (he must have been a Jaguar fan, a *'giaguarista'*). That wasn't good news for our wallets! After we passed the Ponte della Becca, the Jaguar let us overtake him because we knew the way.

The roofer and Cassani weren't just there to inspect the roof; they were also interested in the location and overall condition of the house. Foreigners buying a house in Italy to start a bed and breakfast were a new species to them. The gentlemen were surprised at the conditions inside the house (everything new, fully furnished and all!) and at the beautiful location. We discussed the positions of the new windows, where the new staircase to the cellar would go and what we wanted a terrace for. Cassani, just as at our first meeting, knew how to cut to the chase and wanted to know what other building work we were planning, apart from roof repairs, before we went back home for three months. The windows? That was not possible, because that needed permission from the council, and obtaining one of those could take a couple of months. The terrace? Not possible, for that we needed the permission of the neighbour, whose land was adjacent to the foundations that would need to be laid.

For now, we could only tackle the roofing. Cassani gave us a quick total estimate: twenty five euros per square metre for a roof of approximately hundred and sixty square metres. That works out at four thousand euros. Oops, that's more than we had bargained for. We immediately asked if this was a reasonable price. "You won't be able to do it cheaper," the *ingegnere* assured us. Yeah, how can you possibly check up on that in a foreign country, especially when you really have no choice. The roof was leaking like a sieve and we wanted to make sure it was repaired before our break back home. I thought I had better ask how long it would take and how many labourers would be involved on a project like this. In the Italian building industry, quotes are calculated on the basis of square or cubic metres, not according to man-hours. That's not useful if you want to create your own realistic estimate. The roofer claimed that this job would take a bit less than two weeks with about three men working on it. The roof had to be stripped of all the tiles first if you wanted to make a clean job of it. The wrong way of doing it was demonstrated by the current state of the roof.

Alright, go ahead, we sighed and said goodbye to four thousand euros. Cassani, in his turn, put pressure on the roofer to really get the job finished in early February. "I said *presto* and now I expect *presto*, the beginning of February and I will be coming out to see it for myself!" A real toughie, our *ingegnere*! We gave him our keys so that he could come and measure up everything the following week, and draw up a *progetto* containing all the options together with cost estimates. Making some progress early would mean that the time we were going to spend back home from March to June wouldn't completely go to waste. The phrase *progetto* suddenly made our building project very real. Were we really starting a *progetto*? Were we stretching ourselves too far? Oh well, we'll soon find out, we thought with some spirit, a Dutchman is no *paurista*, pushover.

Tutto a norma

Italy is a country full of rules and regulations, but these rules and regulations were not created to shed light on what is right and what is wrong, in fact quite on the contrary. It seems that they were actually designed to deprive one of clear-cut solutions. Imagine if there were only one simple rule, without complicated conditions or ambiguous definitions. That might have the result of making it impossible for you to do whatever you please. No, the rules must accommodate several possible interpretations. And old laws are not abolished when new ones are created. The ideal situation is when there are several separate rules which contradict each other. Then you can go ahead and have a ball. The best type of legislation is the incomprehensible variety that can keep whole schools of lawyers working in vain forever. It's not for nothing that Italy proudly holds the title of the country with the most lawyers. In Rome alone, the last survey counted sixty thousand lawyers: a complete town!

According to the sales agreement, drawn up by a sworn-in solicitor, the previous owners had provided us with a guarantee that all services (gas, water and electricity) were installed in the house, and complied with current regulations, *tutto a norma*. But you were supposed to understand this loosely, as was explained to us by the electrician we hired. He showed us the bundle of electrical cables tumbling out of one of the fuse boxes. Electrical cables stuck together by duct tape, loose wires, cables of the wrong colour. Plenty of choice. "But it was all according to *a norma*!", we said in surprise. In response, he made a typical Italian gesture which involved pinching the fingers of his right hand together and making a clockwise turn from his wrist. 'Of course it is.' Meaning: 'Do you really think so?'

The first problem that our electrical expert tackled was that of the boiler. We just couldn't turn it on. It was a question of inserting the plug into the socket upside down. "It's alternating current, shouldn't that make a difference?" we asked, taken aback. "Oh," said the electrician, "you are not even supposed to use an external lead to connect the boiler to the electrics in first place. That's not *a norma*." For a couple of hundred euros he could buy the necessary equipment to fix the most serious breaches of the regulations, he told us.

There was a lot more wrong with the boiler. We thought we were being clever, and contacted the plumber who fitted the boiler. We found his name enclosed in the user's manual that we had received from the previous owner. Even though it was Sunday, a day of rest, the day of the Lord, he came out. We, being sceptical Northern Europeans, were surprised that he wanted to come out on a Sunday, and took some extra cash out in case, he was going to charge us his Sunday fees. That was not the case as it happened. This kindest man ever made sure, first of all, that there was gas. The *riduttore di pressione*, the pressure regulator in the metre cabinet, kept on faltering and we couldn't get it to work.

Looking at the gas boiler he regarded the pipes with concern: it looked like the work of a *dilettante*, a DIYer. That's not done: that is totally 'ab-normal' and not *a norma* at all! He could still remember how it was originally connected before it had been clumsily botched up. It was him who did the first installation. He recommended a *ricorsa*, a small re-routing the pipes before making a start on the boiler. "How much is that going to cost us?" we asked nervously. "Oh, that's a couple of hours' work, times my hourly wage," he answered. If we gave him the keys, he could work on it during the week in between his other jobs because he lived very near, at the bottom of the hill in Scorzoletta. "Yes," he admitted, "I was born in this house! I lived here more or less from the very day it was built in the early 60s." That's why he was so quick to come and check out who had bought his birthplace!

Later it turned out that this good man was not the plumber at all, but the plumber's brother-in-law, who liked to earn a little extra by doing some odd jobs here and there. Not *a norma*, because not everyone is allowed to tinker with gas pipes. On top of it all, it seemed that he made a habit of charging us double the price for all the materials that he had bought for us. Our kind electrician had disappeared after his second visit with a couple of hundred euros that he was going to spend on electric sockets, and we never saw him, nor the set of keys we gave him, ever again. That doesn't sound like *a norma* either. But we were getting proficient at the Italian hand gesture. Clockwise turn.

Sembrava un prete

"*Non è una buona persona,*" said the woman sizing me up with a depreciating look. Whilst she spoke, her index finger was gesturing 'no', as if to give further emphasis to her utterance. I had just come across her on the long road towards Ca' Bosco, the large white house built in a beautifully secluded location on top of the hill above the vineyards. It was a private road that led only to Ca' Bosco and functioned as one long and winding driveway. I had pondered many times who could live there. This lady seemed to be the lucky owner. With her seven dogs.

When Saar and I emerged from the vineyards and stepped onto the road, I was alarmed by the dogs, which started barking loudly. I saw that a couple of them were ready to run at us. Saar became stiff and alert and I hesitated. Should I attempt to walk past the woman and her pack of dogs? That was our planned route home after all. Or was it better to avoid risking a dogfight and if necessary make a detour? I saw the woman was shouting something to me, but it was unintelligible above the pandemonium the dogs were making. She made an inviting gesture which I interpreted as: it's OK, it's safe to approach us. I decided to walk in her direction.

She asked me where I came from. My response, that I lived in Spagna and that Francesco was our neighbour, were met with a stony response from her. I didn't dare ask what the basis of her feelings was, and she didn't go into detail. She told me that she lived in the big white house and part of it had been sold to a family from Milan who only used it as a weekend holiday retreat where they came to relax once in a while. In the midst of seven barking dogs, I thought to myself. The woman revealed that she was on her way to the main street where the mobile grocery van was stopping this morning. She was going to do the shopping. Didn't she have a car? That was unfortunate for someone living at the top of a hill.

At the main road, we said goodbye. As I strolled along, her words were playing on my mind. We had already met Francesco a couple of times and he seemed like a nice enough fellow. With his short frame and bulging stomach, his trousers pulled up to the armpits, he looked like the average Italian old man, the sort you see passing the time at village cafés all day. Chatting, playing cards and watching the world go by. An elderly loiterer. Francesco stood out mainly because of his large, ruddy nose and his strident voice. In the evenings you could often hear his raucous laughter when he was sitting outside with his son and seasonal workers. A bit of relaxation after a hard day's work in the vineyards. "*Rharhaarharharha!*" the shrill sound of crows cawing would reverberate in the air. The only thing that could be held against him was maybe that he was involved in some sort of vendetta with his neighbour, Piero Moro, the details of which we were not privy to. Being direct neighbours, they could hardly avoid each other, and this sometimes led to public confrontations; during which it was mainly Piero's voice you could hear screaming and bellowing ("*Bastardo! Vergognati!*") because Francesco always (wisely?) kept his silence.

The vineyards that lay along the full hundred and fifty metres of our boundary line belonged to Francesco. And that meant that we came across him regularly. Our much-longed for panoramic terrace with views over the valley would be supported by pillars standing on the border between Francesco's land and ours. For this to happen, we needed his consent. Officially, in black and white. How were we going to arrange that with a stubborn old farmer like that, who would probably come up with thousand and one objections, one more absurd than the other? You know what old people are like, old Italian people especially, we were thinking to ourselves. Could he even read and write? We were afraid of how he would respond. We would bribe him if we had to, a thousand euros should cover it, because we had our heart set on that terrace. But events took a different turn.

We decided that we would go to Francesco accompanied by an authority figure, someone whose word bore weight, in case that should be called for. Someone who could put things diplomatically, in the Italian way, because our language proficiency was certainly not up to standards. The perfect candidate: our architect Cassani, of course! We weren't paying him for nothing. With some difficulty, we made an appointment with Francesco (that was already challenging enough; we wondered if he understood us? And we him?). Cassani took the opportunity to bring the official documents with him, in the spirit of 'don't put off until tomorrow what you can do today.' Francesco took in Cassani's imposing presence with reverence. That was a good start. The conversation began with some small talk. Or so we believed. But later it slowly emerged that Francesco had an underlying, deliberate strategy to avoid whatever he wished to avoid. At that moment we were as yet unaware of this.

"This area has suffered a lot through asbestos pollution," Francesco started. "Do you know about the number of people who died after working at the asbestos factory in Broni?" Cassani nodded, it was a well-known tragedy that had been playing out in the Oltrepò for the past twenty years. Asbestos cement. Even now, long after the factory had closed, people were becoming ill with cancer. Sometimes, through the years, several members of the same family died of the same cause. Francesco continued with a wistful expression. "You can still find the stuff everywhere in houses and sheds. In the house down there at the bottom of the hill, where my son lives, the ceiling is still probably made of asbestos. The house has undergone some renovation work by a *geometra* from Santa Maria. A villain and a scoundrel." "*Sembrava un prete*. He looked like a vicar," said Francesco and he put his hands piously together. "But in the meantime... After I had paid him the first instalment he never showed his face again. He received a million liras from me without ever having finished the work! And now I am stuck with that ceiling." Could Cassani come and have a quick look to see whether the ceiling had to go or if it was safe to stay. Francesco's strategy suddenly dawned on us.

Our Cassani was, of course, going to inspect the ceiling. And later tonight Francesco was going to read the official statement that he would need to sign for us (or having it read out to him by his son?). Then he would sign it if everything went to plan. Weeks later, we received a bill for the expert advice that Cassani had provided regarding Francesco's asbestos ceiling. A bill that we paid like pious priests, gladly in fact, because the terrace was becoming reality!

II

The Netherlands

MARCH 2008 - MAY 2008

Ritorno in patria

We had to go back to the Netherlands. Nico's sabbatical was over and I would need to finish my studies at the University of Utrecht. But this mature student had, in the meantime, thrown in the towel. The lack of supervision in Pavia combined with the excitement of buying a house led to my decision to give up my studies. The next months would only get more hectic with all the preparations for our final departure to Italy. In the absence of any face-to-face contact, I informed my Italian 'supervisors' of my decision via e-mail. I couldn't see the point in personally meeting them again. A goodbye dinner at the Carceri? I couldn't bear the thought. Only la Nagel responded to my e-mail. *"Va bene, cari saluti"*, in other words: "OK, all the best!" *La professoressa* Chiara didn't even acknowledge my decision. And with that, my university career ended.

The most important task on our priority list while we were in the Netherlands was the selling of our house. And that was not going to be easy because we knew of some other houses in our neighbourhood which had been on the market for ages. We had put our house up for sale as soon as we made the decision to emigrate, but in the couple of months since then, not one viewer turned up. It was touch-and-go whether we could leave within the foreseeable future. At this point we didn't know that the 2008 property crisis was about to kick off. We had no idea what we would do if the sale took a very long time.

Moving to Italy also meant that we would need to import our trusty old Fiat Punto. It had no value in it left whatsoever, owing to its hundred and twentyfour thousand miles on the clock. In Italy we needed immediate access to a car; without it, we couldn't even get home. But importing a car into Italy isn't exactly child's play: before you can receive an Italian registration number there is a lot of paperwork to be filled in. Paperwork that needs to be prepared in advance in the Netherlands: documents that need to be validated and translated into Italian by a certified translator. Then you need to present these documents at the Italian Consulate. If we wanted to complete this process before our departure to Italy, it was advisable to start early. One of the required documents is the so-called 'roadworthiness certificate', similar to an MOT certificate. This is a declaration from the DVLA (Driver and Vehicle Licensing Agency) that sums up the technical criteria that a vehicle complies with. This seemed a bit superfluous for our Italian born and bred 'lady in red' but rules are rules in the Protestant north. You can apply for a 'roadworthiness certificate' on the website of the Dutch DVLA and for 80 euros you get the declaration sent to you by post within 15 days. But we were unlucky. The certificate didn't turn up and the DVLA couldn't be reached by telephone. In the end, we only managed to complete all the paperwork ten days before our departure! Hopefully, the Italian side of the bureaucratic process would work more efficiently, because over there, more paperwork was awaiting us.

Beyond all expectations, we succeeded in selling our house within the three month deadline that we had set. When we realised that the advert we had placed on the Funda property website had attracted no visitors at all, we decided to 'hitch a lift' with an open day organised by the NVM (the Dutch Association of Real Estate Brokers and Real Estate Experts) by putting up a home-made 'Open House' sign in the front garden. We didn't hire an official NVM estate agent, because we weren't keen on paying for any more 'services' of this nature after our experiences in Italy. We were trying to flog our house via the budget 'DIY' method. We had to organise our open day ourselves. We were lucky that our house was in a good location so that all the potential house buyers who were attending the NVM open day walked past it en route. At the start of the open day itself we hadn't yet found a buyer, but during that very same day, our house had already been spotted by our future buyers. After some negotiations we agreed on a good price and a transfer deadline. At the end of May, we were free to depart, exactly as planned! Now all we had to do was empty the house, i.e. clear all the junk and knick-knacks collected lovingly over the past 19 years, the idea being to try to limit the amount of stuff we had to move to Italy.

A lot of the long-forgotten objects we found boxed up in the attic went straight into the skip. We had a couple of large and bulky items of furniture that we managed to sell, one of these being an extremely heavy Dutch designer cabinet. The cabinet's new owner arrived with a trailer, into which we had to try to fit the cabinet without it tipping over. We were wondering how the buyer would get the cabinet up to his apartment, which was on the third floor, without a lift. Luckily we didn't have to be around to see that. Whilst loading the cabinet into the trailer, we temporarily blocked the entrance to the housing estate. It didn't take long before a neighbour emerged, huffing and puffing, his face red with anger, hurling insults at us. What were we thinking of, blocking the road like this? While right behind our block of buildings there was an alternative route which provided access into and out of the housing estate. We were not going to miss the short-fused Dutch temperament in Italy. We had already observed that when it snowed, the neighbours in Montecalvo parked their cars at the top of the hill, on our neighbour Antonio's plot of land, without it ever coming to fisticuffs. That was one way of being neighbourly.

The outcome of all this clearing out and selling off was that in the last couple of weeks we were left staring at a blank wall, the space occupied for years by our cabinet. The wooden floorboards, which had been hidden under the cabinet had a different colour from the rest of the floor in the living room because they had never seen daylight. The rooms were stripped bare. The dining table was sold as well and the lamp that used to hang above it was now dangling, orphaned, in the gaping space. We filled eighteen boxes with the things that we needed to take with us but which we didn't immediately need. The house became emptier and emptier. A house that was gradually morphing into a place that didn't feel like ours anymore. One more week and we too would be gone. It was really going to happen.

Il trasloco

"Hmm, I'm sure that's well over 15 cubic metres," said the professional removal man after he took a quick glance at all the paraphernalia that was standing ready to be moved to Italy. When requesting a quote from this International Removal man, I had had to give an estimate of the volume to be transported, and I had guessed it would be about 15 cubic metres. Where did I get that from? I didn't want to make a guess in a critical situation like this: imagine if we found out on the day that we couldn't fit everything into the removal van! The stress! No, as a professional control freak, I prepared thoroughly by using one of the free Internet websites, designed especially to help calculate volumes for removal purposes. In fact, I visited not one but several of these websites just to be sure. "Fifteen cubic metres," they all declared in unison. Good, I thought. Job done. But the expert eyes of our removal man made a different appraisal: this was a lot more than fifteen cubic metres. Oh dear, we thought, but as long as it fits in the removal van. The van did look enormous, but volume estimation was apparently not one of our strengths. At this point, the removal man hadn't even seen our swimming pool yet!

Who would take a swimming pool to Italy? Us, of course. Since we had nothing better to do, during our break in the Netherlands, we explored the options for buying one of these 'facilities'. That a pool was needed was obvious, because our guests would expect somewhere to cool down whilst holidaying at our luxurious villa. But how much did one of these things cost? From a quick initial investigation in Italy, we discovered that a reasonable-sized concrete pool would cost around twenty thousand euros. That was a big dent in our budget. Surely there were other options? We decided to surf some Dutch websites: getting rid of the language barrier made things a lot simpler. Wow, look at those above-ground pools: they don't need digging in, you just assemble them on the ground! Maybe aesthetically less pleasing, but a lot cheaper and could be a good compromise for the first couple of years. Because who knows? We might not receive any guests at our villa! After a bit more searching: what do we have here? An above ground pool which can also be dug in, available from a swimming pool specialist in Barneveld. We were intrigued and became even more fired up when we realised that the company was in the middle of a big spring sale. We had to pay them a visit. The consequence of our flying visit to the swimming pool sale in the Dutch countryside was now visible in our garage: hundreds of kilos of building material that, according to the swimming pool specialist, could be transformed with ease into a sleek swimming pool measuring 7.3 by 4.9 metres. Even just the box containing a folded up steel rod, which when unfolded would measure nearly twenty-five metres, was nearly impossible to lift. Well, it was a complete swimming pool after all, with all the vital accessories: pump, filter, skimmer, and even solar heaters so that the pool could get to the right temperature extra fast in the hot Italian sun. It was just as well that the swimming pool dealer had not included the (thirty cubic metres of) water as well!

The removal men didn't seem very concerned: they must have come across plenty of strange situations before. They got to work at once, in the spirit of 'we'll see how it goes'. We had nothing left to do but to watch with concern as the van filled up, with still a considerable amount left to load. Luckily we noticed that the removal men were experienced and had a very efficient way of utilising every cubic inch of space. Seeing this gave us some hope. We felt lucky that we hadn't been stupid enough to hire a truck and move everything to Italy ourselves. That seemed to be the cheaper option, but in our case, it would have ended in disaster. We would probably have had to leave half the stuff behind because we weren't, by any measure, as efficient at packing as these removal men. Where would we have put all the things we couldn't fit in? And when would we have picked them up? It was wiser to spend a bit more, money that we had saved anyway by snapping up a deal in the swimming pool sale. Whether that deal was worth it, we would only find out a year later. We had absolutely no idea of the things that were yet to come.

The removers were not only efficient at loading, they were quick too. Within a couple of hours, the house stood empty and there was still some room left in the van. But as the garage door went up, there was a little surprise... another cubic metre. The swimming pool. Oh yes, and the mountain bike which I recently bought second hand so that I could zoom up and down the hills of the Oltrepò. I didn't want to witness the removal men tearing a muscle whilst loading the steel rod, which I imagined was going to involve a lot of cursing and panting. And I decided to find out later whether they could find somewhere to squeeze my bike in. Instead, I went for a pleasant walk with Saar to our favourite local park. The thought that this was our last walk here brought a lump to my throat. Hadn't we had so much fun here with Saar? I thought back to the countless tennis balls she had caught here mid-air, the times she had jumped into the canal for a refreshing dip when she got overheated from all the running. We were going to leave all this behind. The realisation, that this was the final goodbye, was slowly sinking in.

By the time I got back, the door of the removal van was closed and the garage was empty. Everything had fitted in! Though it was a squeeze. The removal man estimated the volume of our cargo at twenty-six cubic metres, on the basis of the maximum capacity of the van, which was twenty-seven cubic metres. Twenty-six! That was nearly double what I had declared. That meant significant extra costs, but luckily we didn't need to leave anything behind. The removal men departed with all our possessions crammed into their van. We wouldn't see them again until, in a couple of days' time, we had safely set up camp in our new villa ourselves. As we got into our good old Fiat Punto and drove out of our street one last time, a couple of neighbours, standing on their doorstep, waved us goodbye. We were taking with us beautiful memories of nearly twenty years' worth of joy and sorrow in this house and estate.

L'arrivo

"Hier stehe ich, ich kann nicht anders - This is my stance, I have no other choice", Luther is supposed to have said to emperor Charles V at the Imperial Diet of Worms, (i.e. an assembly of various estates of the imperial realm held in the city of Worms), the historic city in Germany where we stopped overnight on our way to our new homeland, Italy. As we stood in front of the colossal monument erected in honour of the Protestant church leaders in Worms, the realisation struck us that Luther's utterance rang true for our situation too. We had no choice either: we had transferred ownership of our Dutch house to the new owners, in the presence of a solicitor, and bravely handed in our keys (or so we believed). There was no way back. We left the Netherlands with mixed feelings. We were closing a chapter in our lives, half a lifetime's worth of experiences and the future was uncertain, no matter how much we were looking forward to this new beginning.

We had sought sanctuary in Worms, as the last stop on a long car journey from the solicitor's office in Zeist. After a couple of detours in the Netherlands, in order to avoid Pink Pop Festival traffic, the journey had been uneventful, and by late afternoon we decided to see if we could find shelter in Worms. As soon as we drove into the town, we spotted a hotel that had one available room left *and* it was *hundefreundlich* (dog-friendly). Our Saar needed a roof over her head too, of course. We had some beer at the bar and chatted with the friendly owner, whilst Saar growled softly in the background every time the hotel's own two hairy mongrels came too near. The owner gave us one last piece of advice for the next stretch of our journey: "Take the Gotthard Pass instead of the tunnel, that's only a short detour and you will avoid the queues." We would definitely follow his advice because the seventeen kilometre long tunnel was already known to us, not only as a notorious bottleneck but also as a terrible stink-hole where the temperature could reach 30 degrees Celsius. We would prefer the fresh mountain air of the Alps.

The night turned out to be a sleepless one. Was it our worries about the future that kept us awake? Or was it our guilty conscience for selfishly leaving all our family and friends behind? No, it was the groups of tipsy party-goers who walked past under our window, yodelling, crooning and warbling, deep into the night. We had lain awake till 2 am, yet we got up early the next morning to start the last leg of our journey. Whilst loading the car, Nico was so focused on the job at hand that he didn't even notice that there was something out of the ordinary: the Punto was crammed with helium balloons! Lots of balloons, in all the colours of the rainbow. Well, of course, it was his birthday today and a very special one too, since his spouse (the writer of these words) wanted to mark this event - besides the cheerful decorations, with a present out of all proportion: an Italian villa! Finally, yesterday's melancholy atmosphere faded into the background in light of our new-found cheerfulness.

We were going to celebrate properly once we got 'home'. 'Home', as strange as that sounded. As far as we were concerned, we were at that moment still travelling away from home. It was going to take some time before we would consider our new house our home. The journey went smoothly, and just before the entrance to the Gotthard Tunnel we took the exit to the Pass. As we were climbing straight up into the mountains, ping, the little yellow fuel warning light came on. Oh damn, I was just thinking about filling up with petrol a couple of minutes ago when we were on the *Autobahn*, but I got distracted and forgot all about it. Ooops! The pressing question was: how many petrol stations would there be along the mountain road? The answer would be 'zero', we concluded as we continued on our way up despite the yellow warning light. "Oh well, gravity will take us down the hill," I said level-headedly, a remark that didn't go down too well with the birthday spouse. Yet we were glad that we had followed the advice of the German hotel owner: on the mountain, the air was fresh and clear, and we were treated to the most breathtaking views, which we enjoyed despite the tight squeeze of the narrow road.

The route past Gotthard turned out to be a nearly empty racetrack (with plenty of petrol stations on the way) and by 6pm we were passing Pavia, with the Oltrepò hills beckoning in the distance. Since we had driven this part of the journey many times before, it felt familiar, nearly like coming home. What a fresh, green landscape! Right after Scorzoletta, we spotted our house, a grey bunker sticking out against the green hills of the countryside: a rectangular monstrosity. Had we really bought that? Yes, we really had! The weather was springlike and fresh, the sky virtually cloudless. Soon we were parking in the drive, and opening our front door. Inside there was one more surprise for the birthday spouse: even more balloons and streamers. This surprise had been prepared by our brother and sister-in-law who had given the house a 'test-run' a couple of weeks ago. Luckily they had put a bottle of local *spumante* in the fridge, so we could drink to the birthday boy and to moving in. Moreover, we were doing so in the sun, on the terrace of our own house! It was too warm for our coats.

Wait, what's that rattling in my coat pocket? Oh no: my set of house keys! I had collected the keys from everyone else and handed them all in, but I had forgotten to hand in my own set. It seemed we still had a way back if we got really homesick.

III

Montecalvo Versiggia

June 2008 - July 2009

L'ora legale

As we were sitting in the Internet café, struggling with our blog (trying and failing to upload the story of our move to Italy), my Italian mobile went off. *"Ciao Stef, sono Giorgio,"* said a hesitant voice on the other end of the line. Then there was a momentary silence because I was so surprised, I couldn't immediately answer. *"Siete a Pavia?"* came the question a couple of seconds later, even more hesitantly. Giorgio's voice sounded disappointed. Of course, we were in Pavia because that was the agreement we had made with Giorgio. Why was he calling us about that? Although we were continuing our battle with our blog's website, we still had plenty of time before our meeting at half past eight. Didn't we? And, although the computer clock showed ten to eight, I knew better: my mobile phone showed the correct time, which was ten to seven. Due to the Italians' infamous carelessness, the computers were still on winter time. At least, that's what my irrepressibly judgemental instincts told me. My logic was faulty though, because if my instincts were correct, the computer clock should have shown ten to six.

But now, as Giorgio was timidly trying to make clear that he had been waiting for us for some time now, a small alarm bell started ringing. He was, of course, right! We had got the time wrong... again! Unbelievable! The am/pm drama from the day we signed the sales contract in Pavia was still fresh in my memory. But how we had managed to get involved in this new misunderstanding was still an enigma to me. We immediately ended our Internet session and began to walk quickly towards the Piazza Vittoria, not to make our *amico* wait any longer. As we were walking, I was suddenly struck by the explanation: I only switched my Italian mobile phone back on today. The phone that I bought 9 months ago, especially to use in Italy. Its clock had never been set to summer time! When we left for the Netherlands, it was still winter time, *ora solare*. But since the end of March, it's *ora legale*, summer time. We went through a whole day in the wrong time zone! And that's why we were so hungry an hour ago (it was seven o'clock instead of six as we had thought) in Café Minerva: the nibbles accompanying the *aperitivo* disappeared very quickly!

Luckily, Giorgio was not offended and greeted us affectionately. We spent a nice evening having dinner at Il Cupolone, where we ordered delicious stuffed aubergines and some sort of pasta parcel filled with *radicchio* and we drank our first glass of foaming *bonarda* since our return. "*Oh, mi sei mancato tanto, caro bonarda vivace!*" we shouted with gusto. Giorgio's portion of black *ravioli* turned out to be a bit on the small side. He cleared his plate within a minute and in response to my astonished expression he just said: "*Erano tre*, there were only three!" (*ravioli*). His mournful expression as he said this will stay with me forever. We both donated some of our more generous platefuls of pasta. That evening and for ages afterwards we had a good laugh about the *quasi italiani, quasi clandestini* and *quasi pensionati* that we were now becoming.

In the first few days after our arrival, we couldn't do much because it was a long weekend, a *ponte*: the 2nd of June is a national holiday in Italy when everyone celebrates Republic Day. But after having acclimatised ourselves, we finally came to life. How could we get registered at our new address? Where were the offices of the ASL, the national health insurance organisation? Was the landline already working? Where could we buy a hand-held mower to fight back the burgeoning weed? What was a mower called in Italian anyway? Could we already connect a TV (including aerial and digital box)? We wouldn't want to miss the latest episodes of our favourite Dutch property show: 'I am leaving'. Moreover, we still needed a plumber to install the washing machine and the electrician had to come back and finish his job. We also intended to visit a solicitor for advice on wills and to formalise the inheritance rights on our *immobile*. Oh yes, and the dishwasher was out of order, and oh yes, the swimming pool dealer was going to send us a skimmer but he didn't have our address yet and oh yes, we had to take Saar to the vet for worming tablets and oh yes, oh yes... We were juggling hundreds of balls at the same time which made it hard to achieve anything in a day. Let alone in six months, as we would later find out.

But the biggest question preying on our mind was: had our *ingegnere* finished the building plans while we were in the Netherlands, and had he submitted them for approval or not?

Un grande lavoro

Did we want our new staircase to the cellar to be supported by a steel construction or use a different method? Did we want an open or a closed staircase? Was it going to be made of wood or stone? Did we want the door leading to the cellar to be solid or to have a glass panel? *L'ingegnere* Cassani was firing detailed questions at us at lightning speed. We sat at his desk regretfully, not having expected to have to make so many decisions. We were hoping that our architect would tell us that the town hall, in the person of *architetta* Roberta, had given permission for our construction work to begin, but it seemed to be more complicated than that. The town hall could only give permission for our building project when they knew which builder would do the work and when this builder had signed a DURC, a statement of 'good professional standards'. A builder! We were nowhere near finding one of those.

Cassani had talked to Roberta the day before our meeting with him, which had given him a push to get on with drawing up a *computo metric*, an itemised summary of all the necessary work involved. Well, a push... Actually, we had expected him to have finished the overview by now. We had even rung him up a couple of months ago to encourage him to make some progress, but when the cat's away the mice will play. So, no *computo metrico* to be seen yet. Yet, we were in urgent need of it because all the builders we were going to approach would need it for their quotations and we needed to see the quotations to choose a builder. A builder, who would, we hoped, sign the DURC, which would enable the town hall to give their permission for our building project to begin (but what if they wouldn't sign it for whatever reason? Let's not consider that for the moment). Writing, writing and even more writing. Obtaining planning permission seemed to be a formality but in order to get it we had to have completed all the planning stages. The time we had spent in the Netherlands was, as far as our Italian project was concerned, lost time.

The *computo metrico* is literally an itemised description of all the work that needs to be done, including all the necessary building materials, their amounts and measurements. How many cubic metres of concrete, how many tiles, how much paint, gravel, how many windows, etc... We could barely follow Cassani's fast-flowing Italian, which was peppered with technical jargon. We resolved to sit down at home with a dictionary and go through everything thoroughly. The next time we saw our architect, we would discuss everything we couldn't understand now. We were increasingly concerned about the total costs as we listened to the seemingly endless list of items.

While all these thoughts were whirling around our minds, we heard Cassani rattling on unperturbed, sometimes interrupted by his shady companion who was again positioned behind him on the left. Bidet? Did I hear 'bidet'? Cassani (or his assistant) was suggesting a bidet for the apartment in the basement. That seemed entirely unnecessary to us, expecting nearly exclusively Northern-European guests. I saw an opportunity to make some savings and I was glad that I could finally contribute to the conversation. "Not necessary," I interrupted Cassani's litany. "Not necessary?" he responded with a raised eyebrow. His dubious companion was smirking in the background. "All civilised people use a bidet!" shouted

Cassani indignantly. As a response, I pointed out the absence of this facility in all public Italian toilets, not to mention the toilet seat situation. It's not for nothing that I spent six months studying at Pavia University... with the Italian public toilet system as my main research focus. "Well yes, public toilets, that's a whole different kettle of fish," sighed the architect. "OK, we can always cross it out later," he concluded and carried on with the enumeration.

Cassani was going to try to finish this *computo metrico* by the end of next week. But which contractors would we ask for a quote? Definitely the roofer, because Cassani owed him that much, after the promise he coerced out of him last February. We were not entirely happy with this commitment because the roofer had already proved himself to be a bit of a slacker. Sometimes he quite happily stayed away for a day, which turned a couple of days' work into weeks. The fibs he came out with to cover his tracks were varied yet deceitful, as we discovered one day as we were sitting in the sun, in the midst of the green vines, awaiting his arrival. Cassani called us after a while to let us know that the roofer had rung him to say he couldn't get to us because of 'snowfall' in the hills. Never mind, our suspicions about the roofer could be raised later when we were going to choose between the various quotations. *Architetta* Roberta recommended another contractor from the village of Santa Maria della Versa (this was another contractor she had an obligation to). We also found a builder ourselves, someone who was working on a house in our neighbourhood. In the end, a firm from Milan had also heard that we were looking for builders and we let them join in the bidding. One of the four must be the one, we thought at the time. But we were not taking an outsider from Scorzoletta into account. Or the DURC.

Cassani was ready to wrap up our meeting, but we still needed to get an idea about the time frame of this whole project. Cassani sighed. Making plans is not the favourite occupation of the Italian. Surely he could give us a rough idea. "Six to ten months," he muttered softly. "Starting from September.' We sat dumbfounded in our chairs. Did we understand him correctly? "Yes, yes, this is a *grande lavoro*." And nothing was going to happen before September. In August the whole of Italy is on holiday, and before that everyone is busy finishing current work, explained Cassani. There went our autumn season. And our spring season too if we were not careful.

On our journey back to Montecalvo, a solution was slowly taking shape: wouldn't it be better to delay creating a basement apartment for a year? That was the most costly part of the project and if we left it out we could still start renting out the first apartment from the beginning of next year. Moreover, that would allow us to keep a better grip on costs, we would get an impression of the builder's work ethic and after a year we would be in a better position to decide if we had enough guests to warrant the opening of a second apartment.

Perhaps it was better, to do things like an Italian would, a bit more *piano piano*? You see, we were quick on the uptake, before long we would be Italians through and through. When will the project be finished? Oh, *domani* (tomorrow)!

La residenza

"*Si stanno divertendo un sacco,*" said the friendly clerk to his colleague in the town hall of Montecalvo. "They're having great fun with it." With what? With the omnipresent and unavoidable *codice fiscale*, the Italian equivalent of the national insurance number. The last time, we were told by a town hall clerk that in Italy you could leave the house without money, but that you should never go anywhere without your *codice fiscale*. She had learnt hers by heart which seemed quite impressive since the code is rather longer than a security PIN: the *codice fiscale* is made up of a combination of about twenty numbers and letters. Yet this time she nearly forgot to ask us for our codes. The form that she needed to fill in for us was terrifying: an A3-size sheet, covered in boxes and lines. "I am sure you will need to put the *codice fiscale* in one of those boxes?" I asked with some sarcasm when she said that she had all the details she needed (names, dates of birth, passport numbers, a declaration from the Pavia registry office). "Oh yes, of course! The *codice fiscale!*" she smiled, embarrassed at nearly having made an unforgivable mistake.

Luckily I had had a *codice fiscale* since the previous summer, because the university knew that without it, it was impossible to get anything done. This is why they demanded priority treatment for all Erasmus students from the local *Agenzia delle Entrate*, the Pavia tax office. In principle, it would be easy to construct your own national insurance number, because all the letters and numbers were derived directly from the person's name, date of birth and place of birth. All the characters, except for the last one, which was a random letter. Our surveyor Buttini, had once given us a detailed explanation about how you could construct the code yourself (except for the last digit). There were even Internet pages set up to do the job for you.

Still, it managed to go wrong at the tax office. After we had filled in yet another official *modulo*, form with our details, the administrator behind the window had inserted it into a tube, which took it via an internal postal system to a no doubt secret department within this impregnable tax establishment. After some time had elapsed, the answer had surfaced from the depths: another form containing the vital *codice*. Relieved and elated, we took off home, taking with us this valuable prize. But it didn't take long before we realised that Nico's code wasn't right: one of the letters in his name was spelt wrong. A quick check on an Internet site confirmed our suspicion. We had to go back to the tax office.

The clerk behind the window at the Agenzia didn't blink an eye. Unperturbed, he collected all our details again and inserted the corrected message into the post tube. What would happen next within the mysterious insides of this tax monster? Would they notice that the same person had applied for a second *codice*? And what would they do about it? What would happen to the wrong code? Could it be destroyed? Or would it lead a meaningless life forever in the insides of the 'beast'? We would never find out, because soon a brand new code rolled out of the tube, one that we checked on the spot. Correct!

Armed with our codes, we were now trying to obtain *la residenza*, 'residency', from the town hall in Montecalvo. This was necessary to avoid having to pay a much higher tax rate, because those who are not registered in an

Italian town, have to pay much more tax. The form to fill in was enormous, but the rest of the formalities turned out to be negligible. We already knew from experience that things were different in Pavia. Months and months ago, we had to unearth everything there was to know about the two of us in order to be allowed to register as residents in Pavia: *codice fiscale* of course, passport, insurance certificate, credit card, tenancy agreement, etc. Luckily, thanks to Giorgio's skilful form-filling, we were not suspected terrorists! We had already received a temporary declaration to this effect but we didn't know whether it was the final version: the town hall had never informed us of that. And in order to be able to register in Montecalvo, we needed a declaration of this type.

This was going to be our next enquiry. At the first window in the registry office (where we had already submitted our details months ago) they couldn't help us. "That's not my area, I only prepare the files. I have no idea what the other departments do with them and it's none of my business. You need to go to that other window, on the other side of the lobby," was the defensive response to our friendly enquiry. But the damned 'other' window we hit another dead end, because it was manned by an ignorant and indifferent young bimbo. We exchanged the following dialogue:

"A declaration of residency? Yes, we can do that, you will need to bring in the following documents: *codice fiscale*, passport, tenancy agreement, etc."

"But we have already provided those ages ago!"

"But we still need them."

"Does that mean we are not yet registered?"

At this, the bimbo started typing something on her keyboard until some sort of residency document appeared on her screen.

"Yes, your application has been registered, but in order to get a confirmation, you need to bring in all your documents again."

"But you've already got a file on us which contains all the information you need!" we responded indignantly.

"Oh, really? Those papers need to be handed in at this window and that hasn't happened yet."

"Does this mean that our file couldn't get from one side of the lobby to the other in all this time?"

"I don't know, I don't know," the bimbo answered apathetically.

We realised that we had reached a dead end. How were we ever going to become residents of Montecalvo?

"Are we officially residents in Pavia?" we asked just to be sure, hoping that a verbal confirmation would bring us a bit further.

"Yes, take a look," came the surprising answer. "Look on the screen. Would you like a printout?"

We were speechless! Isn't that what we had been begging for, for the last 15 minutes?

"It will cost one euro and twenty cents per print-out, mind you," added the bimbo as if she could scare us off with this extra piece of information. No, just leave it, it's not worth it! We both got our own copy and went on our way in total confusion, nevertheless with great relief: we had still managed to get the confirmation that we couldn't get! An opportunity not to be missed.

In Montecalvo, luckily everything went smoothly and without any queueing: that's the big advantage of living in a small hamlet. Within an hour we were declared residents of Montecalvo Versiggia! If you want to buy a house in Italy, at least make sure that it's in a small hamlet!

I fratelli

One of the two brothers was sitting behind the counter carefully soldering a chain, whilst the other was chatting with some customers. One of the brothers was called Dante, but the other wasn't Beatrice. They didn't look like each other. One was thin, had greyish-blond stubbly hair and a moustache, the other (with the soldering iron) had sleek dark hair and no moustache. The thin one also had a strange tic: he kept on stretching his shoulders and back as if he was trying to straighten an uncomfortable jacket. But he wasn't wearing a jacket. He had this nervous tic, while his brother was calm personified. The wall was hung full of photos of several generations of Crosignani, spanning a total of forty years of expertise in domestic appliances. *Fratelli Crosignani Elettrodomestici* was an established name in Santa Maria della Versa and its neighbourhood. The next generation was already preparing to pick up the baton, in case the brothers were ready to hand it over: a young nephew often helped out in the shop and specialised in modern gadgets like mobile phones and tablets.

We were no strangers in the brothers' shop. We popped in a couple of times a week for plugs, extension leads (which funnily enough are called *ciabatte* here, slippers), satellite dishes and vacuum cleaner bags. There were always visitors in the shop, but usually not to buy things. The locals often popped in for a chat or to discuss the latest news (about plugs and sockets and such). If a customer happened to wander in, the grey-haired Crosignani brother would interrupt his conversation and address the newcomer while straightening his non-existent jacket. *"Mi dica!* What are you after?"

The brothers were familiar with all brands. In January we bought a cheap vacuum cleaner at BRICO, the Italian equivalent of a DIY shop. But they didn't supply the bags, except for the first one. The first bag was full as soon as we had hoovered the *cantina*. And now what? We visited several shops but every time we drew a blank. But the *fratelli*! They were different! First, they gave us a sample bag to try out. It didn't fit. Next, the Crosignani brought out a vacuum cleaner bag encyclopaedia and started feverishly leafing through it without success. After they had discussed the case among themselves, the brothers decided that Dante would carry on the search that evening. In a couple of days, we got the results: the *fratelli* had identified a supplier and they could now order the bags. They were no quitters, the Crosignani! If something was out there, they knew how to find it.

Since our very first arrival in Montecalvo, we had been shopping for kitchen appliances at the *fratelli* Crosignani. On the floor where our accommodation was, there was no kitchen installed yet, and for the time being we were using the cooker that Colombo had left behind, in the apartment that we were going to rent out. That wasn't always easy. The gas rings were quirky: sometimes the gas wouldn't turn on, other times it wouldn't turn off. It wasn't really safe. Maybe *a norma*, but not our *norma*. We decided that it was worth buying a new cooker, and preferably not through the 'gas fitter' as he had suggested (*"Te lo procuro io*. I will get one for you."), because we knew that would double the price. No, we had to go straight to the Crosignani, our suppliers *di fiducia*! Who else? Ha, ha, we were thinking, we have another excuse to visit them. We haven't been for at least a day. We were getting withdrawal symptoms. As we stepped through the door, we were immediately greeted, *"Mi dica!"* Unfortunately, we only saw white goods in the shop, no cookers. Would the Crosignani disappoint us? No, of course not. The stubbly haired Crosignani brother led us out of the store, around the back to the stockroom behind the shop. There a whole 'hall of fame' of *elettrodomestici* unfolded before our surprised eyes. This was another example of a typical Italian phenomenon we have encountered before: shops that at first sight seem small and unremarkable later turn out to be well-stocked treasure troves. Only, most of the treasure is hidden in a stockroom out of sight. The Crosignani even seemed to have a large choice of cookers, one of which was in the sale, making our decision easier. Delivery and installation were free of charge. The next morning the sturdier of the brothers and his nephew spent a busy half hour in the kitchen, and that day we became the owners of a new cooker. *Di fiducia!*

A couple of days later we decided that Colombo's oven had to go: it was in quite a state, very rusty and if you turned it on, it made the fuses blow. Far too *a norma* for us. Away with it. The Crosignani had another convenient sale on in their stockroom. When we cheekily asked for a *sconto* because we were *i buoni clienti olandesi*, their most faithful Dutch clients, the price was dropped by another sixty Euros. This time we installed the appliance ourselves, and strangely enough the shiny new built-in appliance went quite well with the old fashioned oak kitchen. *A norma*!

A couple of days later we entered the electronic paradise of the Crosignani brothers no fewer than twice! We took Saar with us and Dante's brother asked what our 'lion' was called: *"Come si chiama il leone?"* He was a dog lover and we only just noticed that near the till there was a small photo of a dog on the wall. It was some sort of little memorial for the dog which must have died some time ago. Saar was also very interested in something behind the till and our dear old friend Crosignani had at first believed that the dog was seeking some special attention from him. But then he suddenly realised that one of the 'locals' had brought him a large stick of salami this morning, which he was temporarily storing under the counter. That's what our Saar was after! The Crosignani brother brought the salami out from under the counter and held it out in front of Saar with a big grin on his face: "*Lo vuoi*? Is this what you want?"

We were not here for meat products, we had popped in to buy some sockets. "Which brand would you like?" came the next question. Phew, we knew what we wanted, but we didn't realise that there were different brands. In the afternoon we came back, this time, armed with the brand name. Yes, that brand was in stock. Our hero Crosignani faultlessly identified the right storage box from a wall of identical storage boxes. Unfortunately, there was only one socket left in the box, but our attendant reassured us: they had already placed an order for more, no worries. He showed us how the storage boxes connected together like a Lego construction: you could change the connectors and the plates at will. Now we had to go home and count how many little 'bricks' we needed.

OK, we would come back once more: for the vacuum-cleaner bags and for the electric 'Lego bricks'. Would the order not be there till Tuesday? Oh well, maybe we would find another excuse to visit by Monday. Otherwise, we could always just pop in for a chat and the latest gossip?

Il medico di famiglia

"Ste-pha-nooos Alo-isiooos ... Ma che nomi strani!" said Dr Dezza to me, whilst staring at his computer screen that was displaying my first and middle names which he had typed in with much difficulty. "Don't you have strange names!" Especially Aloysius struck him with its foreignness. "Would that have an Italian equivalent?" *"Luigi,"* I said. *"Ah, Lu-iiigi,"* said Dezza. But he didn't seem convinced. He didn't press me to explain the origins of my surname (*"Smooolders"*). He probably wouldn't be able to derive my *codice fiscale* in one go without mistakes.

But apart from this slight weakness, everything about Dr Dezza radiated sharp-wittedness: his clear twinkling eyes, his fine-set lips, yes, even his thin metal-framed glasses. He was keen to understand everything, just as keen as he was to explain things, in other words to lecture. For no particular reason, he gave us a small lecture about the effects of the medicine he was prescribing: its advantages and disadvantages and how it interacts with the human body. And every time that ironic look in his eyes. As a GP he came across small ailments every day that resulted from the weaknesses of human nature. Forgivable weaknesses, but weaknesses all the same. Years of practice had made him an expert in the human mind and body. The spirit is willing, but the flesh is weak. People living in this part of the world were mainly led into temptation by the wine, the cheese and the pasta. "Don't eat too much cheese!" he said at the end of each consultation.

The first time I wanted to visit Dr Dezza, I was unlucky. I went to the town hall and took my seat in the waiting room again. This time, my business was of a medical rather than an administrative nature. Dezza held two surgeries per week here, but a note on the door said that he would be absent for the next couple of weeks. He held daily surgeries in Santa Maria della Versa, so I decided that it would make more sense to try to see him there. This turned out to be a good decision because he only had access to a computer and a printer in Santa Maria. He needed these to print out his referrals and prescriptions and to register me at his practice.

In order to get a *medico di famiglia* and *di fiducia,* you first need to be registered with the ASL, the Azienda Sanitaria Locale, the Italian national health insurance, and to be in possession of a *la tessera sanitaria,* the insurance certificate. For this, you need your *residenza*, residency declaration, and the unavoidable *codice fiscale*. Luckily we had already taken our first steps on the slippery ice of bureaucracy and hadn't fallen over yet. As a genuine Montecalvo Versiggia resident I had the right to a *medico* and I could take my pick out of four GPs serving in the district. The GP had to agree to take the new patient on and both parties could refuse further service if there was a case of *la turbativa del rapporto di fiducia,* in other words: if trust had been broken. I took a chance on Dr Dezza in the absence of any references.

On my first visit to Dezza, I took a folder containing my medical records which I had obtained from my Dutch GP. I did this to inform my new GP of my medical history and all my ailments. Except that the documents were all in Dutch. Dezza didn't seem bothered and leafed through the folder with deep interest, glasses on forehead, nose nearly touching the paper. He could understand most of the medical terms because of their Latin names. He muttered and chuckled to himself, oh yes, the weakness of the flesh. Let me see, back problems, hernia, diclofenac, ... "The back problems are getting worse," I complained, "Twenty-five years ago, all I had to do was swim once a week and that helped, but now that doesn't seem to be enough anymore. Why is that?" Dezza had a twinkle in his eyes and was smiling. "The difference between now and twenty-five years ago is that you were twenty-five years younger then." But we might as well take an X-ray of course. He was also going to refer me to a physiotherapist, but I mustn't think (as everyone else seemed to think, according to Dezza) that my back would be back to normal after ten sessions. I would have to keep up with the exercises at home, of course. I nodded seriously but I saw that Dezza had no *fiducia* in it. "Did you see the old man who was just leaving?" he asked me. Yes, I had seen him. He had seemed like a spry little fellow. "Do you know how old he is?" No, maybe seventy or so, I thought. "Ninety years old!" Dezza told me, almost proudly. "He has just returned from a trip to Sardinia, on his own! He drove to Genoa and took the ferry across." Dezza was gloating. That such a thing was even possible, he seemed to be thinking. But the punch line of Dezza's anecdote was yet to come: "He came to complain. Things aren't so easy anymore, he's getting faster than he used to. He wanted to know what was wrong with him! Ninety years old!" Dezza shook his head.

I visited him regularly because of my back, and in the end, managed to get that physiotherapist referral. Every time I came to see him, he searched for my folder and displayed it on the computer screen ("*Alo-isioos*," Dezza shook his head). When I started up about my back again, he laughed and said: "*La vostra schiena ormai è un mito!* The story of your back is taking epic proportions!" Smile. Twinkly eyes. Did I make the right decision when choosing this *medico di famiglia*? For the time being I didn't see any reasons to declare a broken trust between me and my *medico*: we were far from a *turbativa* yet. I did have *fiducia* in Dr Dezza.

Il meridiano

If you turn right out of our gate, the road takes you to the tiny hamlet of Spagna. If you turn left, you will find our neighbour Francesco's large and relatively new house. Opposite this house stands his former home, now occupied by his unmarried son, Roberto. In the summer, the house also accommodates seasonal workers from Romania and Albania. Beyond this house the little asphalt road becomes a wide gravel road that winds through across the vineyards down to the valley and joins a main road. Behind Francesco's villa there is a little row of houses, of which the middle two are owned by some Milanese people who only visit once in a while in the summer. The last house in the row is occupied by Francesco's daughter and her husband. The first house bordering Francesco's house is the odd one out in this neighbourhood: it belongs to the *antennista* Piero. Francesco calls him *Il Meridiano,* the Southerner, because Piero originally comes from the South of Italy and for many Italians from the North that's tantamount to a foreign country.

Piero and Francesco were in a vendetta, the ins and outs of which we were not privy to, although Piero had wasted no time warning us of our untrustworthy neighbour, as he thought of Francesco. The woman with the seven dogs from Ca' Bosco was of the same opinion and warned me that Francesco was no *buona persona*. When Nico last came across Piero in the vineyards, whilst out on a walk with Saar, Piero told him that he was an *antennista*, an installer of TV aerials. Just the person we needed. He promised to come over soon to see what we needed doing and how he could help us. But once inside our house, at the kitchen table, he was brimming with revenge against Francesco and seemed to want to talk only about him, no matter how we tried to steer the conversation back to satellite dishes.

"I want to warn you, please don't be offended if I seem rude, but I see that you are kind and trustworthy people and I don't want you to suffer the same fate that I have. I see it as my moral duty to warn you. Be careful with that charlatan; don't do business with him because he will rip you off. I have lost ten thousand euros. He was going to do everything for me but nothing came of it. He also stole a piece of my land: according to the land register, my garden is a lot bigger." Piero was just rattling on, going around in circles and coming back to the same topic every time. *"Nico, Stef: ascoltatemi, non fidatevi, ricordatevi quello che vi ha detto Piero. Vi faranno male. Ve lo dico dal cuore."* With that, he placed his hand dramatically on his chest. What were we supposed to make of all this, as newcomers and outsiders?

We had already heard a variety of bad stories about Francesco but apart from his clever trick, roping in our architect Cassani, there wasn't much we could resent him for. Or was there? Francesco had once asked us urgently to fence our boundary off with a couple of sticks and some wire. "You need to do that because if someone drives across your land with their tractor a couple of times they might start to see it as their right to do so, and you lose control over your land," said Francesco. We didn't think twice about it, and anyway, we wanted to stay on his good side, so we erected a wobbly bit of chicken wire. We only found out much later that Francesco's advice wasn't given entirely selflessly: he had a different agenda, while keeping us in the dark.

Considering Francesco's reputation, we didn't really trust him. But Piero wasn't a very personable character either. Filled with hatred and revenge, he wasn't really able to show any interest in others: he didn't listen. His small back garden (which even he didn't dare to enter), was home to four bloodthirsty dogs which barked day and night, terrorising the neighbourhood. Not a good demonstration of thoughtfulness towards others. Had he got them to scare Francesco away? When Piero finally arrived to install the satellite dish, after several follow-up visits which he spent bad-mouthing Francesco, he behaved like a dictator towards his son who came along to help him.

During the first months since our arrival in Spagna, we witnessed many strange scenes which we watched in surprise. We could hear a public argument between the two neighbours from our balcony, Piero shouting: "*Vergognati, vergognati*! Be ashamed! Be ashamed!" What Francesco should be ashamed of, was not clear to us. We couldn't hear Francesco. A couple of weeks later, we saw to our astonishment that a barrier had been erected behind the little row of houses. A real, official, red and white railway barrier. Was it there to stop Piero from accessing the street on which the houses stood? At the beginning, Piero lived together with a girlfriend, who suddenly disappeared from the scene, from one day to the next. Soon after this, we heard that Piero wasn't working for the aerial company anymore, he was no longer an *antennista*. How did he earn his living now? Later on, he ran for mayor of Montecalvo, only to suffer a humiliating defeat by the incumbent.

Another couple of years later, during which time we tried to avoid Piero as much as possible (every time we met him in the street he would start up again about

Francesco), we saw a removal van appear in the street. He couldn't, could he? But it was true. To our relief, Piero departed, taking with him his noisy pack of beasts. He had sold his house to another of those Milanese families who would only visit occasionally at weekends. The railway barrier remained as a testament to the old days. "Remember when the *meridiano* was still around?"

I Due Padroni

We had to give our future B&B a catchy name, not something commonplace like Bella Vista or Panoramica. When you looked those up in Google, you couldn't see the wood for the trees. No, it had to be a unique name, something that would immediately say something about the owners too. After some doomed attempts, it finally dawned on us how we could be best described: we were two men with a dog who accompanied us everywhere. We are in fact two masters, of the B&B and of the dog, in Italian: *I Due Padroni*. The fact that we were also two *padrini*, godfathers to our nephew, we thought better not to mention, especially since *padrino* is the title of the most famous *mafia* movie ever: *The Godfather*! In order to avoid any mistaken association with the movie, we decided to attach the pretentious-sounding 'Villa' to our name, hoping that after the refurbishment, our grey concrete cube would indeed be transformed into a luxurious residence. This is how the name 'Villa I Due Padroni' was born. At least in theory. Now for the harsh reality.

We had to make sure that the well-heeled tourist with an interest in Italy could find us on the Internet. Or even better: that they couldn't avoid us. We had to create a website that would contain lots of information about ourselves, our house and the area. A website that would easily come up on a Google search. While one *padrone* was constructing one beautiful web page after another, applying website knowledge gained through years of self-study, the other *padrone* went off in search of suitable tourist information for our future guests. Unfortunately, there wasn't much information available in any language other than Italian, not even in English. We would need to make our own translations if we wanted our visitors to find their way around here.

During the search, we stumbled across lots of small wineries, lots and lots of wineries. These vineyards, each created their own special wine variety: *bonarda, buttafuoco, croatina, sangue di guida, pinot nero, riesling,* etc. All the producers gave their wine their own, fantasy names, and that's how I came across a wine called Ca' Padroni, a *buttafuoco* from a wine maker whose company name was the pretty-sounding Il Piccolo Bacco dei Quaroni. That was funny: could this be our house wine to give our guests as a welcome present? We should make sure we tasted it. Not long after this discovery, I saw an advert from the same vineyard organising a Sunday open day among the vines, where participants would have the opportunity to pick and press grapes. The pressing was of course done in the old-fashioned way, with bare feet! A couple of months later you could buy your self-made (*bonarda*) wine which would be named '*Questo vino l'ho fatto io*'. I have made this wine myself. This was the opportunity we were waiting for!

On the appointed Sunday we set off to pick grapes. Unfortunately, the day was cloudy with the prospect of rain despite months of sunshine. But we set forth. We arrived on the dot of 10 o'clock at the hidden little winery, *puntuale*, as Northeners do, but the Italians were nowhere to be seen. The ten to twelve other participants arrived about an hour to an hour and a half later, from Milan and its suburbs. We had enough time to have a nice little chat with Laura and Mario the owners of Piccolo Bacco, over a cup of coffee and some almond cake *fatta in casa*. This was followed by a tour of the winery. Wine production requires surprisingly small amount of space: the machinery would easily fit in the back garden and doesn't seem very complicated (but things might be different if you want to make really good quality wine). Mario told me that he used to be a stockbroker but a couple of years ago, he and his wife had decided to leave the rat race behind and to start up this business. He made the wine, his wife was in charge of the restaurant.

As the rain began pouring down, we and the other participants (who had all arrived in the meantime) climbed up onto a cart that was pulled along by a tractor at a snail's pace. We headed into the vineyard. Armed with crates and secateurs, we got to work. The task turned out to be more difficult than expected: the bunches hang back-breakingly low for us tall Northerners, and in the tangle of leaves and sprigs it was hard to find the tiny stem of the grape bunch that you were trying to pick. We had barely filled our crate half-full with tight bunches of bright purple grapes (our crate was surprisingly heavy already), when the activity was called off: the Italians amongst us had decided that that was quite enough. We, *padroni* through and through, looked up in surprise: we had only just started! Italians and rain... The Italians took their time over lunch, a good two and a half hours, but the menu didn't help, there was too much to choose from. We have earned a good lunch, haven't we? With all that hard work! All the dishes were home made, *alla casalinga*, and the delicious sparkling *bonarda* that accompanied it was home produced.

After lunch, it was finally time for the much-anticipated grape-trampling the most enjoyable part of the programme, nice and messy. The children were especially quick to exploit this opportunity, but us two *padroni* also ensured we were part of the action! Socks and shoes were removed and we stepped, without hesitation, into the large red bowls. It felt quite pleasant underfoot, like a massage, even though today the sweet juice was a bit on the cold side. At the end of the day, you could design and order the labels for your own bottles. With your signature on. Or something else. Something else? How about a photo of our villa? Yes, that was possible. This is how we found the wine that we would welcome our guests with: *Questo vino l'abbiamo fatto noi: I Due Padroni*!

It was only on a later visit that we got to taste the real Ca' Padroni wine. A wine of this type, *buttafuoco*, is protected and can only be made from grapes grown within a certain area and also has to comply with a number of other strict guidelines. It was quite a strong, harsh wine, not to everybody's taste. It was the sort of wine that you would need to 'learn to drink'. Not the right house wine for our holiday apartments because the *padroni* would never be harsh to their guests, they like to keep things simple.

Pronto soccorso

An Italian car pulled into our drive. Who could it be? It was late afternoon and we were sitting on the terrace at the front of the house. Two men got out, a driver and a passenger. The passenger was Dutch and he immediately launched into a monologue without introducing himself or stating his business. He assumed that we would know who he was because he merrily chatted away about the countryside and his adventures on the way here. "I knew your address approximately and I figured I would find my way here with a little local help. But I couldn't find any information about Montecalvo in Piacenza. Luckily this kind gentleman gave me a lift even though I was further away from this place than I expected." Oh, OK, we were thinking, so what did you come here for? When he saw our frowning faces, the penny dropped. "I'm here to pick up the car, on behalf of the Royal Automobile Club," he explained. Finally, everything became clear. Of course! The car that Nico's brother and sister-in-law had to leave behind after the *infortuno*, Nico's brother's accident.

A couple of days earlier we were sitting with them on the same terrace, enjoying our aperitifs when we suddenly heard a bang followed by a hissing sound. Nico ran indoors because he was convinced that something happened to our dinner that was cooking on the gas hob. The rest of us – Nico's brother and his wife and I – stayed in our seats, thinking along the same lines: Nico would deal with it. He came back after a couple of minutes with a very different explanation: the supply pipe of the sink in the bathroom which had been – very annoyingly – leaking for a while and had seemed irrepairable, had now come off, and the bathroom was flooded with water. Brother-in-law and sister-in-law got stuck in with the mopping up and repair work. I went to the *cantina* to look for a new pipe fitting. As I was walking back, I heard some stumbling, shuffling and groaning and caught some fragments of the conversation, like "Yes, no, careful, no, like that, yes, that's better." What was the matter? I went inside and saw Nico and his sister-in-law carefully lifting his brother onto a kitchen chair. He couldn't straighten his leg: his knee had locked up.

We decided that there was no other option but to go to the First Aid Station. Where was it actually? We just had to look it up. It seemed it was a 15-minute drive away in Stradella, while the car was also a 15-minute stair-climb away. We came to the conclusion that the villa was definitely not wheelchair friendly. On the way to the *Pronto Soccorso,* we studied the dictionaries we had hastily brought with us to find out how to explain to the doctor what had happened. The knee is *ginocchio*, locked up *bloccato*, etc... Trying hard to memorise all the words and phrases. We arrived at the first aid station safely and our patient was placed in a wheelchair by two burly nurses and rolled away through an open doorway to a treatment room in the back. We followed, armed with a dictionary. Or maybe we didn't follow because one of the two nurses pushed us out and pulled the door shut in our faces. *"Non è permesso,"* we heard her say. "No access!" There went the helpless patient with a crooked leg and no Italian language knowledge whatsoever. That was going to involve a lot of arm and leg waving, we thought, but then again, maybe not leg waving. We had nothing else to do but sit and wait and hope for a happy ending. How good or bad was Italian health care anyway? We would now find out thanks to our brother-in-law who had sacrificed himself entirely selflessly for this experiment. Luckily, no sounds of sawing or screaming penetrated the waiting room. It didn't sound like there was going to be an amputation. Through a slit in the window, we could still follow the proceedings to some extent.

After about twenty minutes the door swung open and our beloved brother-in-law reappeared, still in a wheelchair but now with a bandaged and somewhat straighter leg. The nurse had succeeded in straightening the leg without using too much brute force, but that was as far as it went. The knee refused to cooperate any further, and the conclusion was drawn that this meant a torn meniscus. This needed an operation. Oops. Now we really had to think out a plan B. We had to ask the insurance people to arrange a flight back home because driving was out of the question. But in order to be allowed to fly in this state, the patient had to take anticoagulant medicine and he had to demonstrate that he could walk independently (even if using crutches). In order to get our hands on the medicine and the crutches, we needed a doctor. It was, of course, a Sunday. The weekend doctor! How does that work in Italy, we wondered?

It went very smoothly: the doctor on duty appeared within a very short time at our isolated location. Although, he did seem a bit eccentric. He looked like he had all the time in the world and as if this was a purely theoretical problem that you could philosophise about for hours on end. We, on the other hand, just wanted a prescription for the anticoagulant and some crutches. Today if possible, not *domani*. "You could undergo the operation here in Italy," suggested our medical house-philosopher airily. "I happen to know a really good orthopaedic surgeon (*di fiducia*?) who specialises in knees. He is the mayor of Varzi, you know, the little medieval town in this region."

But the combination of the middle ages and health care didn't really appeal to us, and the patient was even less impressed. And what to make of a surgeon mayor? It sounds a bit like the Singing Butcher, all the more because Varzi is well-known for its delicious *salami*. Yet to have the operation in Italy had its advantages: according to our medic, you would be back on your feet within a day, you would be able to walk, true, on crutches, but you would have your mobility back. They would be able to continue their holiday in Italy. Going back straight to the Netherlands would mean being placed on a waiting list for weeks (months?), which didn't sound good. But is that really true, the quick-fix operation? The insurance company wouldn't have any of it and insisted on a return home. So be it. Once back in the Netherlands, after a couple of weeks on the waiting list, he underwent his first operation. The first one, because it wasn't successful and he had to have more surgery (after some more orderly waiting on a waiting list). All in all, our brother-in-law spent a year in and out of hospital. It does make one wonder: could it have gone differently? Perhaps the medieval surgeon wouldn't have been such a bad choice after all.

So their car was left behind; it would be collected by the pick-up service provided by the Royal Automobile Club, who seem to send their volunteers (because it's volunteers who do these sort of jobs) on their way without a map. But all's well that ends well. In the end. At least we had the chance to get some valuable training in the provision of *pronto soccorso*. You never know when this knowledge might come in handy whilst running a B&B.

Torti

No, he didn't want any *grappa* and no wine either. He was *astemio*, totally teetotal, he claimed. We stared at Torti in surprise. Alcohol abstinence in this part of the world? You must have great self-discipline. Later I did catch him pushing a trolley at the Cantina Sociale, the wine merchant where local drinkers replenish their fast disappearing wine stocks for two and a half euros per bottle. What would he be doing there? Buying a bottle of *grappa* for his daughter, was his excuse. A bottle, with a trolley? Subsequently, he pretended to know which wine would suit which meal and which vineyards produced the best wines and which were rubbish. But this wine know-how probably originated from his stubborn nature, which we would get to know through and through. Torti always knew better.

We had a meeting with Torti and our architect Cassani in our kitchen to pick a definite start date for the building work on our villa. In spite of all the dawdling and delays, finally the day had come: we were going to sign the contract with the contractor, Mauriziano Torti from Pietra de' Giorgi. Why Torti? Because he would never turn up drunk on the site? No, that's not the only reason. We asked for several quotations from various builders. One of them didn't even reply (the builder that *architetta* Roberta recommended through the council), another one only managed to produce a note with a sum (Cassani's roofer) and the third one produced a quotation without ever coming out to assess the site of the building project (our candidate from Santa Maria). Only the Milanese builder took the trouble to come out and inspect the house and ask additional questions. He also produced the only sensible quote, for a sum of money that seemed reasonable. The choice seemed to be easy.

But we didn't take our neighbour Francesco into account. He came over with some friends to discuss our building plans and that's how Torti's name was mentioned. Torti lived in the hamlet of Scorzoletta within the county of Pietra de' Giorgi, a settlement in the valley of Scuropasso stream. Our house faced the valley, and the hamlet was visible from our window. The local pub, owned by our *idraulico*, plumber, was near enough opposite Torti's house. This was the *idraulico* who was born in our house and therefore just loved helping us out. Or was it because he doubled the price of everything he had bought for us? We had had quite enough of him when he asked 600 euros for a pump that cost 300 euros on the Internet. Colombo had explained to us that the waste water was pumped from the holiday apartment's kitchen through a waste tank to a septic tank higher on the hill. When we located the hole in the ground where the waste tank and pump were supposed to be (according to Colombo), it was empty. The waste water was simply seeping into the ground! We had to install a pump, and that would be just the job for the plumber. He seemed to know where you could 'get hold of' the best pumps, for a mere 600 euros, he said. When we confronted him with the fact that you could get the same pump on the Internet for 300 euros, he simply corrected himself: "600 euros? No, I said 300." His electrician mate, also from Scorzoletta, had in the meantime done a runner with a couple of hundred euros, but our 'doubler' said that he would have a little chat about this with him and that he would make the electrician bring us the things we had already paid for. He would come *domani*. Really.

Torti had also slithered out of the same snake pit and at first we weren't very keen to ask him for a quotation. That would only cause further delays and we really wanted to get to work. But Cassani convinced us. Torti had paid him a visit (behind our backs) and made a really good impression. Or had he bribed our architect? We were becoming more and more paranoid in this country. Who could you trust, who was really *di fiducia*? Torti had already worked on previous extensions to our house and his familiarity with the property might be of great benefit, said Cassani. He also promised to prepare a quotation as soon as he had assessed the current state of affairs. "OK, let him come over then. But he will need to come immediately and make his offer on the spot," we said to our architect. Torti came indeed very quickly and he promised to bring back his *preventivo domani*. But the next day there was no Torti and no quotation. But luckily, in Italy, every *domani* is followed by a new *domani*!

It took until the end of the week before we saw Torti again. Mumbling some incomprehensible excuse, he handed over a crumpled brown envelope, the contents of which we read with impatience. The quote was a bit higher than that of the Milanese builder but it had more detail in it. It was undeniable that it looked good, and had the advantage of Torti's familiarity with the house and the fact that he lived so close. To top it all off, he was a teetotaller too. But again, that little hamlet of Scorzoletta, did seem like one big *mafia* family. The *antennista* had warned us about Francesco who had defrauded him with his workmen friends. Torti was recommended to us by Francesco, the builder from Scorzoletta. What were we getting ourselves into? Could we really trust Torti the 'teetotaller'? We tried to cling on to Cassani's best judgement, in the hope that his integrity wasn't compromised by the influence of the Scorzoletta gang.

Cassani speed-read the proposal at the kitchen table. Torti watched him with indifference. After a while Cassani got to the section on the *sconto*, the reduction of the final sum, that is often offered at the signing of a contract. He threw a questioning look at Torti. He muttered that he had had some bad experiences with price reductions, but that he would consider it (*domani*). Cassani didn't pursue the subject and asked about payments: were these always made on the basis of an invoice? Yes, he always gave an invoice, declared Torti stony-faced. At this point Cassani winked at us, which seemed to mean: we will make sure of that. We signed the contract and drank a toast. Torti didn't, not even a glass of water, no matter how *di fiducia* its source was.

Cowboy Land

We were stuck. Truly stuck. Sideways on the road. Groups of cyclists kept on coming round the corner at high speed and being scared to death by the car blocking the road, which could only be avoided by braking hard and steering even harder. Rude Italian expletives were thrown at our heads. There was a deep ditch behind us leaving us no room to turn around. But driving forwards would have resulted in the car completely blocking the road, which could have proved fatal because we couldn't see if any more cyclists were coming. Yet it wasn't an option to remain as we were, sideways on the road. We had to take a gamble.

On our way to an agility competition for dogs in Voghera, we had unwittingly stumbled into the middle of a cycle race. We seemed to be going in the opposite direction from the race, climbing up the winding road uphill, and every time we came across a group of cyclists on the descent, we were as shocked as they were. It was an awkward situation and there was a point when we just couldn't go on any further. We decided to turn round, and that's how we got stranded sideways on the road.

We risked it and completed the three-point turn. We were relieved that none of the tour contestants had ended up under the wheels of our Fiat Punto. We took a different, less risky route. We were covered in cold sweat nearly all the way to Cowboy Land, the amusement park where all kinds of clever dogs were going to demonstrate their acrobatic skills. We had already had our share of stunts for the day.

As we arrived, we immediately struck lucky because the agility competition for medium sized dogs, *taglia media*, was just about to start. Cowboy Land is a bit like a theme park with a 'Wild West' vibe, where you can stroke goats, look at bison, throw horseshoes and learn to lasso. We collected all the information we could get our hands on for our future guests with children: unfortunately, the attraction is only open at the weekend and there are no information leaflets in English or any other language. "No, we never get any foreign visitors," said the lady at the ticket office. I'm sure you don't, and at this rate, you never will!

It was good fun to watch the clever dogs jumping enthusiastically over bar jumps, doing the dog walk, creeping through tunnels and slaloming around poles on a very tight course. It was even more fun to watch their masters hurrying breathlessly after them. They all did pretty well, the dogs were well trained. Some of them overdid the barking a bit as they were completing the obstacle course. They didn't quite get what the big deal was: "OK master, I will jump over that fence again if you like. Don't you worry yourself," they seemed to be saying. The championship was won by a Norwegian. We saw him walking up and down with his fast brown furry friend just before the beginning of the race. It was a Dutchman who told us who the winner was. He was attending with a whole group, all taking part in the world championship. Some of the spectators were holding orange banners, one of them displaying the full text of the Dutch national anthem. You never know who you meet! "*È piccolo il mondo,*" as the Italians would say.

After the race in the medium sized dog category, it was time for the race for the disabled category. Disabled dogs, we were thinking. What's that going to be like? But we were wrong, it wasn't the dogs who were disabled but rather their owners. First, there was a warm-up round so that they (i.e.: the masters) could familiarise themselves with the course. The dogs were watching with amusement from the side. Watching the warm-up round stirred my innate sarcasm and evoked the memory of a Monty Python sketch: the 'Ministry of Silly Walks'. During the race, the masters were clearly a delaying factor: the dogs often had to stop and wait. We soon realised that the real competition had taken place during the registration process for the race: how could you get registered with a disability that hardly made any impact on the speed at which you could walk around the course? A Czech man won by completing the course at a speed that would have been acceptable in the 'able' category. What was his disability? We wondered.

At one point, a father had lifted his child out of her wheelchair and carried her around the course. As they went, the child was waving her little arms about, giving instructions to their dog who was walking beside them. Did the child move her arms independently, or was the father helping? Was it a real child or a doll? We watched with astonishment. All the participants were moved to tears. We, on the other hand, felt mainly embarrassment at witnessing this display. As they concluded the course the spectators gave them an ovation. Is this a cultural difference? Are we Northerners too sober or are we so suspicious of false emotions that we cannot allow ourselves to be genuinely moved?

After the race, we took the opportunity to indulge ourselves in a rare fast food feast. After all, we were in an amusement park. Chips with, of course, hot dogs. No squished-cyclist sausage meat for lunch. Thank God.

Via Matris

The church in Scorzoletta didn't look particularly striking. Maybe because it wasn't a historic building or a modern concrete design. No, Don Luigi's church was simply built of brick. What we did find strange, however, was that there was a pile of assorted building materials in the churchyard because there was no sign of construction work on the building itself. And why was there a van parked next to the church with an unforgettable but mostly impractically long name written on its side: www.santuariodellamadonnadelperpetuosoccorso.it?
Was the church a company with a website? And what were all the signs surrounding the church? There was even a sign to the nearest defibrillator in case you happened to be struck down with cardiac arrhythmia as you were walking past! Sometimes we playfully called the parish the Madonna of the Pronto Soccorso (First Aid) but we didn't expect anyone to take it this seriously.

The only traffic lights in the whole hamlet had appeared only recently on the road leading to the church. When they turn red, it's anyone's guess why. The ways of God are inscrutable as are those of Don Luigi. Because all these strange activities had been initiated by the uncrowned ruler and priest of the parish of Madonna del Perpetuo Soccorso. He has been running this place with an iron fist for the past thirty years, engaging and supervising everyone involved in his numerous projects. When you step through the beautiful glass doors of the church, doors that wouldn't look out of place in a Dolce & Gabbana store, you are dazzled by the marble and gold interior. This *is* a Dolce & Gabbana store! Dazzling modern mosaics decorate the wall behind the choir and the rest of the inside looks perfect too. The parish seems to be swimming in money. Or is there another explanation?

Indeed there is. In a country like Italy, where the Catholic church still commands respect, and in a small village where social hierarchies still persist, Don Luigi has succeeded in getting all the villagers to support him in his holy duties. Contractors deliver their goods for free and hope that in return their soul will ascend a marble staircase to Heaven. Luigi is selling redemption and accepting payment in kind. When you get to know him better, you realise that he is a truly omnipotent busybody, who interferes in other people's business and coerces them to listen to his opinions. According to hearsay, 30 years ago, before he became the priest of Scorzoletta, Don Luigi used to be a priest in a church near Canevino. That was a special church because it was in the same area as one of the most famous regional producers of *spumante*, the champagne of the region. It seems that the young priest found it hard to resist the temptation of the sparkling nectar and that he was often found in a heightened state even before evening fell.

Now that Don was well into his seventies, age had started to take its toll. During mass, Don sometimes read out all the instructions on the hymn sheet, such as: "And now the priest will break the bread and the congregation will answer," which made the mass last twice as long. The holy water just stood there on the side in a plastic washing up bowl instead of a beautiful font and Don would wave at it indifferently when he read out the instruction which told him to bless the holy water.

Until recently, Don had shared his home with a woman, but the nature of their relationship was unclear: a so-called housekeeper. During mass, she always stood in the doorway of the sacristy, in full view of the congregation, invariably wearing a dirty looking bandage on her ankle. While the choir was singing, she kept her unblinking eyes turned heavenwards, looking like the blissful embodiment of Mary herself. But one day she was gone. She was said to have passed away suddenly, and the villagers couldn't get enough of this news. "*Poverina*! The poor lamb!" said the women to each other. Don Luigi ruled on seemingly unaffected and undisturbed. "And now the priest will remember all those who have passed away in this Parish."

One of Don's most recent initiatives was the organisation of a procession which he christened Via Matris: Our Lady's Path, a variation on the more familiar Calvary. It was an exercise in crowd management. It took some talent to organise the chaotic gang from Scorzoletta. On the evening of the procession, everyone gathered in the square in front of the *cantina*, the wine merchant on the outskirts of the village. After a short period of confusion during which the organisers completed the final checks (Where was the missing sash? Where were the matches to light the lanterns with? Who would be walking at the front?), the cavalcade had set off, led by Don Luigi. The procession travelled down a walkway that was specially laid down for this occasion, a unique feature in this region.

The image of Mary was carried on a four-wheel drive Fiat Panda at the front which had been 'donated' for the occasion by a local wine maker. The procession stopped at seven locations, and the parish choir attempted to sing a verse from the Stabat Mater at each stop: the music score was virtually illegible in the dark and the piece itself was not well rehearsed. One of the stops was at our *idraulico*'s café (we wondered if he charged Don Luigi double as well?) This was followed by a stop at the wine cellars of the Count of Vistarino (the earthly ruler of the place), one at the church and another one at the restaurant Savini. That's how it carried on, past several wine makers, the whole route comprising about eight hundred metres, which seemed to be more than the Fiat could take. The car broke down halfway ("And now the priest remembers the car that was sacrificed during Via Matris.") and had to be pushed along by some of the sturdiest parishioners.

Via Matris ended with a small party, involving dancing village girls, an old accordion and a genuine *spettacolo pirotecnico*. Fireworks! The three tubes of Bengal Fire from Lidl lit by the priest on the edge of the walkway, produced mainly smoke, and only enough sparkle to be visible from the other side of the street.

Il DURC

"*Pronto!*" is what Italians say when they answer the phone: "I am ready, go ahead." But how do you, as the caller, know that you are talking to the right person? Italian wireless communication is not prepared for questions like that. The idea is that you immediately get to the point without knowing for sure if you are talking to the right person. The word *pronto* is an often used word in Italy. It's a shame that when it comes to practical business, it takes a lot longer to get it *pronto*.

Although delayed, as expected, we now had a detailed project plan and a builder, Torti. Now all we needed was for the council's *architetta* to consent to our plans, which would bring Montecalvo Versiggia a significant economic *boost*. But not so fast. Without it ever having been mentioned before, it seemed that we were missing another document. Sometimes it seemed like everyone we did business with, the *architetta*, the *ingegnere*, the builder, was doing their first ever building project. We needed a DURC. A what? A *Documento Unico di Regolarità Contributiva*. It appeared that a builder had to declare professional standards for every new job (or job phase), no matter how small, and present it at three bureaucratic institutions: the INPS, the INAIL and the *Casse Edili*. Professional standards meant that the builder declared that they paid their taxes and their compulsory national insurance, that they respected the interest of their client and that they had satisfactory third-party liability insurance. This paperwork was passed between all three official bodies within one month. After a month had passed, the building project was allowed to start, whether the DURC had been completed or not. "Torti applied for the permission ages ago, didn't he?" we asked our experienced architect - our tower of strength - looking at him with a mixture of hope and anxiety. Of course, Torti and Cassani (with their substantial collective intelligence) had already thought of that. Hadn't they? Dear Cassani, please have mercy on us, we pleaded silently. No such luck.

The damned DURC took exactly one month to complete. A month in which the weather was only getting worse and during which Torti kept on encouraging us to make a start. Go ahead, the DURC will sort itself out, he said. But *architetta* Roberta had warned us: if someone fell off the scaffolding, Torti was not insured, which would make the employers liable. We had no choice but to refuse Torti's offer every time, albeit with a heavy heart. By now it was raining nearly every day and each time we saw Torti he was shaking his broad head gloomily. If we could just... No, Torti, no, not today. Not without the DURC.

By the beginning of October the DURC had arrived: an unreadable document folder with three very important stamps. Torti had passed. He soon arrived with a truck, all kinds of digging and demolishing equipment, and his team: an Italian *muratore*, a Moroccan handyman, an Albanian labourer and a Romanian giant. Later, once we had befriended them - we were united against our common enemy - we learnt that the Romanian and the Albanian, as *extracomunitari* (non EU citizens), were not declared by Torti and he paid them cash in hand. Torti also expected the *muratore* to do heavy lifting work, despite the fact that he was aware that he had a heart condition. Using protection against falling off the scaffolding was unheard of. Helmets? Industrial ear muffs? Torti was in possession of the official stamps, that took care of everything. He could carry on doing things the way he always had done. No wonder that every couple of months, Italians wake up to news stories about horrendous building site accidents, which often claim lives. All the papers and TV channels dissect the ins and outs of the accident for days. Opposition politicians demand action and the politicians in power promise immediate measures. New, stronger legislation is what is needed! But we knew from experience that it had nothing to do with the old or new legislation. The average Italian citizen doesn't give a toss about the government and will only fulfil the minimum official requirements so that they can carry on unhindered and follow their own stubborn ways. Italians and authority...

Oh well, we were covered, we had the required stamps. Our building adventure had begun, we thought enthusiastically, entirely unaware of everything that was yet to come.

Una corsa della morte

One day I had to visit the ASL to hand in some forms which would help finalise our insurance against all conceivable medical costs due to illness, accident or other misfortune. The ASL, *Assistenza Sanitaria Locale*, (to be pronounced as *Asluh* and not as we first thought, *Asul*), had a branch in the village of Santa Maria della Versa which was open a couple of mornings a week. This was very useful because otherwise you would have to travel 15 kilometres to the bureaucratic stronghold of Stradella, where you would lose the will to live being faced with their ticketed queueing system. Santa Maria on the other hand is only a 45 minute walk away. But today I decided that I would test run the mountain bike I got in the Netherlands. It had taken the professional removal men a lot of effort to squeeze the bike into the overflowing removal van: I had to make it worthwhile.

Until today, it had stood there lonely and neglected in the big, empty *cantina* of our house. In the Netherlands I had only ridden it once. It was a pretty uncomfortable ride because I wasn't used to mountain bikes. Soon, in Italy, when I use it every day, I will get used to it, I had thought. And now we were both ready for action, Mister Mountain Bike and *me*. Well, nearly ready. I still had to fix the toe clips onto the pedals: imagine if one of your feet slipped off! It could send you into a fatal crash or you could get your privates crushed on the crossbar! Good thing I'd thought of it in time! The tyres felt firm, I could set off. I walked up the first slope, from our gate to the country road, because it had a gradient of 14% (!) and I hadn't mastered the gear system yet. I didn't yet know that changing gears is something you just do instinctively.

Once I got to the top, I got on the saddle of my sporty two-wheeler. I picked a gear at random and it turned out to be more or less the right one. Mountain biking! Anyone can do it! After a bit of effort uphill, it felt good to be flying down the slope. A fresh breeze caressed my athletic body. I let the bike free-wheel to the bottom and was prepared to apply the brakes because I hadn't become a stunt man just yet. I squeezed hard on the hand brakes, left and right, but nothing happened. Nothing! The brake on the back wheel didn't respond either. The first reaction that overpowered me was that of total disbelief: this was not possible! There was a funny side to it too, as if I was playing part in a film. Father Ted, on his way to his first sermon, face-plants on the gravel path: "Dammit!" There is nothing wrong with the brakes, I told myself, still in denial. I squeezed one more time hard on the hand brakes, staring in disbelief at the brake-pads. No response. How was that possible?

My initial disbelief quickly turned into panic. As the adrenaline coursed through my veins, I considered all the options which might help bring me to a stand still. Steer the bike into the grass bank, I thought at first, to prevent it going even faster and to limit the damage caused by a fall. The prospect of the impact made me hesitate. In the meantime, my speed kept on increasing. As did my panic. I had to make a decision now! But just as I was ready to scream out in fear and head towards the grass bank with my eyes shut, I saw the way out: a bit further on, there was a side road sloping upwards. I had to go up there! I knew that at my current speed, if I missed this chance, I would never be able to take the hairpin bend, that I knew would soon be coming up. If I cut through that corner, I would fly off a steep hill and land on the banks of the fresh little stream of Versiggia about 20 metres below.

The side road which I now absolutely had to take was near the entrance of a transport company that traded in sand and gravel. Huge, heavily loaded trucks came and went from the site all day long. If one of those lorries was just leaving as I was approaching the side road... Don't think about it, don't think about it! Completely rigid with fear, I steered the bike towards the side road and after about fifty metres uphill I came to a stop. And that's how it ended, as quickly as it had started. At one moment you are staring death in the face, and at the next you carry on living as if nothing had happened. Well, maybe not as easy as that, because later on the day I was hit by a wall of fatigue.

I only discovered later how this nearly fatal accident could have happened: our friendly removal men had to remove the front wheel of the bike when loading it into the van. Once in Italy, they had neatly fitted it back on. I wasn't aware of any of this and I found the bike in Italy in the same condition as I last saw it in the Netherlands. At least that's what I thought. But I didn't realise that when they removed the front wheel, they had also disconnected the brake cables. Reconnecting them would have taken a couple of minutes, but the removal men hadn't given it that much thought. Accidents can hide in the most unexpected places. How ironic that this had to happen on the day when I was going to sort out my illness and accident cover!

To make things worse, when I arrived at the ASL office (this time driving to Santa Maria in the car) I found it closed. Outside, the opening times were displayed: Tuesdays and Thursdays 11a.m. to noon. But if you peered through the glass door, you could see another note, saying that the office was open between 9a.m. and 11a.m. I was one 'near-death' experience richer, but still not insured.

La chimica

Suddenly we heard shouting coming from the *cantina*, the cellar. "*La chimica! La chimica!*" Had Torti's builders stumbled upon an environmental time bomb, some deadly poison in the ground? Were they having an argument and threatening each other or Torti with chemical weapons? No, they were talking about some sort of paste which they were going to use to stick the metal frames holding the door and window frames to the concrete wall. In the absence of a better term for this paste, the labourers and Torti referred to this stuff as *la chimica*.

The first couple of days of the construction work stirred up quite a bit of dust. Building work starts with demolishing. This is often the quickest bit because you can do it without thinking. Demolishing? Anyone can do that, even unwittingly. Holes had to be opened, a lot of holes. We agreed with Torti that he should start on the top floor, where we were going to have our living quarters. We would live downstairs until all the work was done. Then we would move upstairs, and Torti could carry on hacking away downstairs. It was a perfect and comforting plan, which would completely fail in reality, as we would find out later. It was clear from the start that staying in the small live-in kitchen of the hire apartment wouldn't be fun, and in practice, it was even harder. We were surrounded by noise all day long, and slowly but surely dust crept in everywhere, even though we had sealed all the doors with plastic. The labourers' comings and goings made it a very chaotic environment to be in, and what's more, Torti would call us over every few minutes to make decisions about every little detail. At the end of each working day, we collapsed exhausted on the sofa, in the sealed, yet dusty living area of the upstairs apartment and attempted to take in whatever was on TV, without much success.

Upstairs, the builders first made a hole for the new kitchen window, followed by an even bigger hole in the wall of the bedroom and another hole of the same size in the study. Only the bathroom and the living room were untouched for now, but in the case of the bathroom, peace was only temporary. We had already hired another demolisher who would extract the remarkable little child-sized bath tub from the shower area and create a sleek modern walk-in shower cabin. If we're demolishing stuff, we might just as well do it all at once, we had enthused at the time.

Before the demolition work had begun, Torti had warned us that it would take three weeks before the holes would be fixed again. That was three times longer than his earlier promises, so I made a little joke to Torti that it had better be ready before the first snowfall. He laughed and made an Italian gesture which meant: "Go fool someone else!"

In addition to the top floor, they started to demolish things outside and also in the cellar. The twenty metre long concrete wall along our boundary with Francesco's land had to go. It was leaning too much to be used as part of the supporting structure for our panoramic terrace. Our Romanian giant and his little mate Festi, the Albanian labourer, hacked away for days on end, using a deafening hammer drill (without ear muffs); and they kept taking away dozens of (too heavy) wheelbarrows filled with rubble. In the *cantina*, a redundant and hideous blue bathroom had undergone the same fate (demolished!), and a new entrance opening was made. The enormous industrial door would be replaced by a normal entrance door. That meant: demolition work! We were woken every day of the week (apart from Sunday) around 7a.m. by the arrival of the builders: roaring engines, hooting horns, chatting and, if Torti was there, loud shouting and swearing (we weren't wearing ear muffs either). He loved to give clear instructions or at least clearly audible instructions. The louder, the clearer, was his motto. Some days, when he turned up late, we noticed how much better the work was progressing. Because Torti interfered with everything and everyone, he knew everything better by definition and kept treading on everyone's toes. He also liked to 'help' us regularly with good advice: how could you tell when your *borlotti* beans were ready, what's the best way to prune an oak. When he came back later, he would quietly condemn the job you had done or shake his head in disapproval. "No, this was the work of an amateur. *Non sei capace*. You don't know what you are doing." But I wasn't impressed by the Italian's pruning skills since I had witnessed the damage that Pavia's gardeners could cause, once they got hold of a chainsaw. In a couple of minutes, they could turn the most beautiful trees into sad skeletons: they sawed through the thickest branches without mercy, without giving a thought to the result for the shape of the tree. Topiary? Never heard of it. Once they had finished, it was like looking at post-modern art: an open air exhibition inspired by the Dutch artist Zadkine's Second World War monument in Rotterdam entitled "The Shattered City". Bare, thick branches reached up to the skies like amputated arm stumps.

During the day, Torti often left the site for an hour, much to our and the labourers' relief. When he returned in his truck or grey Punto, you could see him coming from far down the valley. "*È lui, è lui,*" shouted Festi, and everyone hurriedly left their smoking or chatting station to be diligently engaged in an activity by the time Torti was pulling into the drive. We also interrupted our pottering and fled inside, into the live-in kitchen, which was soon, inevitably, penetrated by the foreman's bellowing.

È piccolo il mondo!

I had been trying to get through to the restaurant the whole evening, without any luck. If there was no answer even at 8 o'clock in the evening, then they must be closed. Azienda Agrituristica Bagarellum was only open at weekends anyway, and booking was required according to their website. I was determined to visit this establishment because their restaurant was highly praised on the Il Mangione website. This usually very critical Italian gastronome (who grew up on *la mamma*'s delicacies) enjoys writing about the restaurants he has visited and leaving his reviews on the website. Bagarellum was, according to the website, one of the best loved kitchens in the whole area, just as good as the excellent but much pricier Vecchia Pavia Il Mulino in Certosa di Pavia. Moreover, when we found out that Bagarellum was located in Montecalvo Versiggia, close to a walking route which we had once done, we became determined: we had to try this restaurant. Even if it was for the sake of our visitors!

In the end it took quite a while before I took action and picked up the phone to reserve a table. Nico was out performing with the parish choir of Scorzoletta, and after the performance they would be going out for pizza. I decided to ring now to reserve a table for Sunday lunchtime. This *pranzo della domenica* is something quite special in Italy. Whole families go out in their Sunday best and sit at the table from about 1p.m. until late afternoon, sampling a variety of dishes. First in line is the *antipasti*, the appetiser, which could consist of regional cold meats, preserved vegetables and onions, a *frittata*, omelette, or whatever else is in season. This is followed by two *primi*, pasta or *risotto* (perfectly *al dente*) with fresh vegetables, meat and cheese. Then two *secondi*, meat dishes, and finally the dessert, *il dolce,* which might be rich chocolate cake or *crostata* (fruit tart). Luckily there is always plentiful home-made wine at hand to wash it all down. And if you are lucky, to aid your digestion (Italians are always mindful of their *salute,* health) you will be offered a delicious *digestivo,* such as a *nocino,* walnut liqueur, or a self-distilled (hence illegal) *grappa,* one of Italy's most popular spirits.

We were glad to sacrifice ourselves to such treatment once in a while. But unfortunately, I couldn't get through to the restaurant. At about half past ten, I made one last attempt, and struck lucky: they picked up the phone! "*Pronto*!" I couldn't make out clearly what the owner was saying because of the noise in the background. That sounded good: it sounded like they had a full house! I asked in my best Italian and at Torti's volume if I could reserve a table for lunch, for the following day, Sunday. "Yes, you can. Tomorrow you need to be here at 1 o'clock," said the woman. Saar was allowed to come too. "What name is it?" asked the friendly voice. "Smulders," I said and started spelling it, because to an Italian ear this name means nothing. "*Esse, emme, oo,...*" "Yes, that's enough," she said. "It's fine." Well done, I thought, feeling very pleased with myself. When Nico arrived home late in the evening from his meal out with his fellow choristers, in a rowdy pizzeria in Arena Po, I told him proudly that I had booked a table at the most sought after restaurant in our district.

On Sunday morning we left the house at around midday. The morning fog had burnt off and the weather was pleasant, perfect for walking. The route took us on a steep climb; our destination was hundred and fifty metres higher than where we had started. *Frazione* Bagarello was a charming dainty village with little old houses and little old people, free ranging chickens and other loose farmyard animals. A centuries-old farming village (the first written mention of the hamlet of Bagarellum dates from the 12th century!), nestled in the midst of vineyards with breathtaking views all around! And how peaceful too!

Once we had admired the village and its views, we headed towards the entrance of the restaurant. But even before we went in, two people came running out of the kitchen, straight towards us, shouting "Nico, Nico!" "What are you doing here?" they asked. "*Mangiare*, eating," I answered in surprise, but Nico had already recognised them. They were Nando and Leda, two choir members who owned this restaurant. The restaurant that I had booked yesterday while they were sitting at the table in the noisy pizzeria, right next to Nico. "Yes, I noticed that Leda received a business call that involved two foreigners," said Nico surprised. "'*Prenotazione per due stranieri*,' I heard Leda say. But I wasn't aware that they were restaurant owners and that you were on the other end of the line..." Nando and Leda shouted out in astonishment: "*È piccolo il mondo!* It's a small world!"

After this big surprise, we got everything we had hoped for: the lunch, which lasted for hours was incredibly tasty. All the dishes were prepared with home produced ingredients, Nando had told us. And you could taste it! It was the first time that we had *risotto* with truffles, that not only tasted but also smelled of *tartufo*; instead of the kind that we had had elsewhere, that cost 90 euros and tasted like cardboard. We also had some delectable pumpkin-filled pasta with *funghi porcini* sauce, made with wild mushrooms. And the food smelled divine! The delicious aromas stimulated our sense of smell and we relished the meal even before we took our first bite. The secret behind this success was eighty years of cooking expertise in the family. Over the years, *la nonna*, Leda's grandmother had slowly instilled the mastery of cooking into her granddaughter. It was clear: as far as we were concerned, none of our guests would go home until they had had a meal at Bagarellum!

Ciak ciak

After a week of demolition work, Torti's mood began to lift remarkably. He was cracking jokes and stopped harassing us continuously with his unsolicited advice and disapproval. Instead, he spent most of his time sitting in his dilapidated truck, which we nick named 'Dumpy', busying himself with paperwork. He was preparing his first invoice because our architect Cassani had declared that enough work had been completed to merit some payment. Weren't things going a bit too fast? We wondered, a lot less enthusiastically, afraid that our bank account couldn't keep up. Every aspect of the construction work was broken down and agreed on in the contract. Cassani was in the best position to judge how the work was progressing because he prepared the *computo metrico*. He had come to the conclusion that the labourers had (already!) clocked up some overtime: instead of demolishing a part of the boundary wall, they had demolished practically the whole wall. This cheered up our builder even more.

Torti concentrated hard on his calculations and when he needed to take a break, instead of dishing out criticism, he came over for some friendly gossip. For example, about our house's previous owner, Colombo. Last week the verdict had been delivered in a lawsuit between Torti and Colombo about the payment for the work that Torti had done on the house, eight years ago. "*E tutto per una trave!* All that fuss over a supporting beam!" snorted Torti. It was obvious that he could still wind himself up about it. He explained that after the building work had finished, the living room wall had developed mould.

Colombo made a complaint about this. But Torti was convinced that the mould was Colombo's own fault because he had put the heating on without having aired the house thoroughly first. That's what you get if you do that. Torti thus refused to treat the mouldy wall, to which Colombo responded by refusing to pay for the damned supporting beam. Torti threatened with a lawsuit and Colombo met his challenge. Colombo had lost the lawsuit last week and now he owed Torti ten thousand euros. "For a supporting beam that's worth 100 euros at most!" smirked Torti. "And Colombo was already up to his neck in debt!" Even the sum he sold this house for would not be enough for Colombo to pay all his debts off, Torti informed us gleefully. The reason: his fickle daughter who had already had three children by as many fathers. Children who all needed to be taken care of. The men had all beaten it, one after the other, gone. The last one had left when he was told that he was expected to work in the *lavanderia*, the launderette that was built in the *cantina*. Colombo was broke. Torti relished telling this story. Even the washing machines were confiscated by the suppliers because Colombo never paid for them. Torti didn't seem to have any doubts about receiving his money.

One morning he beckoned us with his characteristic gesture: swinging an arm towards you whilst making a quick grasping movement with the fingers towards him. At first we didn't understand what he meant but when Torti saw that we didn't respond, he added *"Vieni, vieni!* Come, come!". This is how you trained a dog. Torti summoned us to come. *"Andiam' nel camion,"* he said. "We are going in the truck." Where to? To his house in Scorzoletta, to admire his home, meet his mother and his wife. Our relationship with Torti was suddenly moving on to a new level! We got into 'Dumpy' and it started up, shaking and spluttering. On the way, Torti pointed out all the houses he had built or had worked on in the past. They all looked good on the outside, we had to admit. But, for all we knew, on the inside they could have been mouldy caverns, missing a supporting beam here and there.

Soon, we were sitting in the Torti living room sipping at our coffee after we had admired the dogs, the hens, the chickens and the rabbits. Torti had told us earlier that he loved dogs. He had five dogs, he said at the time, and we imagined a cosy house full of furry scallywags rolling over on their backs. But in reality, the dogs seemed to spend day and night outdoors, and they made a deafening racket as soon as we approached. "My husband adores his beasties," said Mrs Torti with a serious expression, and added: "*È un lavoro, però*. It's a lot of work, you know." Slowly, bit by bit, it emerged that the beasties, the kitchen garden and the whole kit and caboodle were Mr Torti's hobbies ("He never rests, he's always on the go," his wife sighed), but in the meantime she seemed to do all the work. "Am I glad he doesn't like milk! Otherwise I'd have to keep some cows too!" she said, laughing dolefully. Their daughter, who had a degree in engineering but preferred to hang out at home, had a passion for cacti. Mrs Torti couldn't understand why, all those spines. "They do get lovely flowers on them?" we said encouragingly. "Yes, the flowers are beautiful. *Durano poco, però*. But short-lived."

To us, Torti's wife appeared to be the melancholy victim in this relationship. Living with a man whose stubborn and dictatorial nature must have been hard to put up with. But as we were sitting around the living room table, enjoying our coffee, she suddenly lashed out like a frustrated lioness in captivity. Torti launched into giving us instructions about the preparation of *gnocchi*, the traditional potato dish of the area. According to him, it was all very simple: you prepared the dough, which you rolled into little balls and using your thumb, pressed them into the right shape. "Like this. C*iak ciak.*" He repeated the *ciak ciak* a couple of times, whilst demonstrating the hand movement above the table, until his wife suddenly interrupted him: "He knows exactly how it's done. But only from the cook book. He's never cooked a meal in this house in his life."

Il collaudo

"*Che peccato!* That's a shame." The friendly young man at the *Motorizzazione* (Vehicle Licensing Agency) looked at us regretfully. He would have loved to be able to help us complete the extensive amount of paperwork but there was nothing he could do. It was exactly 6.78 euros that should have been transferred to Pavia's Traffic department, not 5.45 euros as we had done. He could only accept our transfer certificate when we had made up the remaining 1.33 euros. Paying in cash? No, that was not possible. There was no other option but a trip to the post office, which was quite some distance away. The Motorizzazione was only open from 10a.m. until 12 noon. We weren't going get this done today.

What a shame! We were already making some good progress with the help of this nice man. We had arrived at the Motorizzazione about a quarter of an hour ago to finish the paperwork importing our good old Fiat Punto. As we stepped inside the building, we were overwhelmed by chaos: there were numerous ticket windows and queues in the big hall on the ground floor. Where were we supposed to be, in God's name? Notices were dotted around on A4 sheets of paper but we couldn't decipher them. But wait, over there by the staircase there was a sign that seemed to say that for *importazione di veicoli* you had to be on the first floor. Did we dare sneak upstairs without permission, not having queued up at a downstairs window first? The alternative was possibly an hour-long wait at the wrong window. We decided to risk it. The worst that could happen would be being identified as unauthorised intruders and getting expelled from the higher offices of this administrative stronghold.

Upstairs there was only one room with the door open, so we decided to take a peek inside. We were in exactly the right place! Our helpful young man was sitting there in the middle of an enormous room, surrounded by tables, documents and files. He immediately understood the purpose of our visit and looked through the pile of documents we had brought. We thought we had done our homework well. We had studied the relevant websites of the Motorizzazione, the ACI (Automobile Club d'Italia) and the Dutch Consulate for hours. We didn't want to get hopelessly lost in the bureaucratic forests of the institutions that concerned themselves with the importation of cars from abroad. We were hopeful that we had everything that was required. The complete translation of the vehicle registration document (all 3 parts) and the declaration of 'road worthiness' which we had received from the Dutch Vehicle Licensing Agency. Everything translated by a sworn translator. Accompanied by several carefully completed forms, downloaded from Italian websites. Plus proof of payment to all the authorities: 5.45 euros to the county, 9 euros to the Department of Transport and 37.79 euros to God knows whose Department of Finance. But as luck would have it, the required payment to the county had just gone up and it hadn't been updated on the website yet.

We were going to have to come back one more time (at least) and we had to hope and pray that none of the other requirements would change in the meantime. Just to be on the safe side, our friendly assistant checked through all our documents. "Was the car tested at a Dutch garage?" Yes, no, well," we stammered. The MOT had expired just before our departure and we didn't think it was worth it to have an expensive Dutch MOT done for our last month in the Netherlands. "You will need to have a *collaudo* done. Without it, you won't be issued a registration number." said our expert. "It takes no time at all, but you will need to make an appointment and take the car to the test centre at Bereguardo." It was going to take longer than we had hoped. Luckily, in Italy you can drive around with a foreign number plate for six months, and we had started our import procedure as soon as we had arrived. Our young man encouraged us to take heart and assured us that on our next visit he would finish all the paperwork in five minutes. "Just come straight upstairs, and I will be here." You got nowhere in this bureaucratic jungle, without a helping hand *di fiducia*.

Transferring another 1.33 euro to the county was no problem at all. What worried us was the Italian MOT. How rigorous would it be? It turned out that they only had an appointment for us in a couple of weeks' time. If it carried on like this, our six months would soon be up. When the appointed day of the MOT finally arrived, we drove to the test centre, hidden in a derelict area on the outskirts of Pavia. Once all the paperwork was filled in, the testing could finally begin. We were holding our breath. What if they found something? What if they needed to make some repairs? Or if we had to come back for a retest in a couple of weeks? The tester lifted our faithful Punto's bonnet. It was an Italian make, maybe that would help. With a serious expression, the tester bent over the engine. At least there was an engine! That's a good start! He wrote up a number and asked us to put on the headlights. They worked: we scored some more points!

It carried on like this until the highlight of the afternoon: the driving test! Nico sat in the driver's seat. His task was to drive a couple of metres and then put on the hand brake. He did as he was told and the car stopped! To our great astonishment, this completed the test: our Punto had passed! The bill came to fifteen euros. One thing is clear: Italian road traffic safety is high on the government's agenda. Or maybe they know that road traffic safety isn't determined by the condition of the car as much as by the quality of the driver?

With our road worthiness certificate and proof of the extra payment to the county, we could return to the *Motorizzazione*, where our assistant *di fiducia* did indeed made short work of our paperwork. In exchange for our folder of forms, we received a small note, which we still had to hand in at a window. At the window, we finally received our new registration number: DP-377-VF. The number plates were immediately ready to take home. On the number plates, there was a little 'i' for Italy in front of the first letters of the registration number: 'i DP' for: *i Due Padroni*! It couldn't get any better! We could even find some meaning for the last two letters: *Vecchi Fannuloni*, old fools.

Was our Punto now completely naturalised? No! To finalise proceedings, the registration number had to be recorded at the ACI, the Italian Vehicle Licensing Agency. If you neglected to do so, you were risking a large fine, the Agency website had warned us. This meant another trip to a different office, with different forms and different payments. This time, we were asked for an *imposta di bollo*, a (tax) stamp that you could buy at any *tabaccheria*, to the value of a strange amount: 14.62 euros. Having said any *tabaccheria*: we did visit quite a few *tabaccherie* to find the stamp for the right value: 14.62 euros (not 14.61, not 14.63, no, exactly 14.62 euros). We did it in the end. We proudly stuck the stamp into the little stamp-sized box on our form because we wanted to make sure not to lose this hard-earned treasure. Now, back to the ACI. After a long wait it was our turn at the Sportello Telematico, the 'telematic' window, which was proudly advertised on the website. Was that a good sign? (Or a bad one?) We handed over our paperwork. The lady behind the window barely glanced at it and said in a dull tone: "The *bollo* is on the wrong form." Oh no, dammit, we thought. How is that possible? We did everything so meticulously and still it wasn't good enough. The stamp was properly stuck down. Perhaps the woman was wrong. She didn't look like the brightest match in the box. She didn't even get what we wanted in the first instance. Only when we pointed out to her that we were required by law to have our registration number recorded did she take us seriously. She was going to ask around for help... from all her colleagues. Without success. But luckily there was a last resort. She went to a small room which seemed to be occupied by the uncrowned queen of the ACI. She was kept busy, of course (by colleagues such as this one), but we were allowed an audience with her because Italian bureaucrats had a soft spot for foreigners. We were going to hold on to our Dutch accents for the time being. The form was wrong, she agreed, but the *bollo* was not yet lost. All that was required was a bit of cutting and sticking. This was the ACI's latest, hyper-modern telematic office after all! We also had to fill in another form. Well, we could barely call it a form. It was a faded A4 sheet, that was a copy of a copy of a copy of an... original, which probably had a statement printed on it in crystal clear black and white, but which by now, was barely legible. Luckily the queen was there with an explanation. I had to declare that I wanted to have my registration number registered. That was what it said. In a nutshell. Name, registration number, chassis number and horsepower (59 KW), signature and done.

Done? No, not quite. This was only our request, we had to take it back to the woman at the telematic window. This was the place where the ACI did its magic using the most up to date technology! Or maybe not? Behind the window there was a lot of writing (more forms to fill in), copying, asking for help (first from every colleague, then straight to the queen), stapling and filing. It all seemed fine. Now we just had to pay the registration fees: 299 euros! That was more than all the fees we had paid so far for importing the car put together. I got out my Italian debit card. At the sight of my card, our clerk's flickering light went out completely: here at the *Sportello Telematico* you could not pay by card for any goods or services you received, no matter how big or small. Luckily we had just - but literally just - enough cash with us. We got a receipt and a certificate of registry and... we were not free to leave. All the whole paperwork had to go back to be seen by the queen. That turned into another half an hour wait in the hall. We watched in solidarity, all the other frustrated Italians swearing inwardly, as they were also spending most of their afternoon there.

Out of boredom, I sent a quick text to Giorgio, summarising our experiences so far. He replied shortly, pointing out the fact that it was us who chose to live in the homeland of Berlusconi and that all Italians knew that ACI stood for *Affari Complicati Italiani*. Let's wait a bit longer. At half past twelve the queen would need to make an appearance if she didn't want to miss her lunch break. And she did. In all her glory. She called the names of all those whining and pining; ours too. We received a magnificent document, much like a diploma. *I Due Padroni sono sopravvissuti all'ACI*! The *Due Padroni* had passed the trials of the ACI.

Il maltempo

Bang! Bang! Bang! *Porca miseria*, will that banging ever stop? I was woken up in the middle of the night by a continuous loud clatter against the bedroom wall. And I had just been having a good dream about Torti... The wind was already blowing hard before we went to bed, but now the shutters were rattling off their hinges. I thought we had made sure they were all shut and securely fastened? Or was it the loose down-pipe that was now dancing in the wind, clattering against the north face of the house?

It wasn't advisable to go and investigate in the pitch dark because of the big pit that had been excavated near there in the previous two weeks. Moreover, the storm was really raging. Now that I was awake, I started to think about Torti. Over how I would grab him by the collar tomorrow and tell him that he really must close up the three open holes in the wall a lot more tightly at the end of each workday. Because, despite all our previous pleas and Torti's and his mates' good intentions, we had found some water damage. And where else, of course, but in the freshly painted and decorated bedroom where we were now trying to sleep. In one of the few rooms of our house that Torti had nothing to do with and which we could safely decorate. Or so we believed. But yesterday, a big wet patch had appeared on the beautifully painted wall, which came as a shock. The rain had probably trickled along the beam that was recently inserted into the wall and had seeped down the ceiling. As a temporary measure, we had covered the beam with a thick sheet of foil, but it would be a better solution to make all the holes in the walls watertight.

Luckily it wasn't raining, because the combination of rain and wind could have been really catastrophic at this point. It seemed obvious to me that Torti had to take some immediate action to avoid any more water damage. But how was I going to get him to do anything? A stubborn old man like that was going to resist advice and come up with excuses. What should I say as counter argument? My thoughts were were spinning round and round in my head. Bang! There went the down pipe again. What was the time? Half past five. By half past seven, the sky was getting lighter and my bedfellow was also woken by the din. He went outside to investigate. The down pipe was not the culprit. The clattering was coming from a small window in the *cantina* that the builders had left open. Thanks a lot!

Soon we heard Torti arriving and I jumped to my feet to grab him by the neck at once. Good job too, because he was only dropping by to remove his tools to safety from the storm. Going up the scaffolding didn't seem like a great idea in these gales: you only had to lose your balance once and a gust of wind would push you over the edge, DURC or no DURC. They had just started on the supporting beams for the balconies, working on the top of the scaffolding. That wasn't a job for these conditions. Luckily, my night-time anxieties were unfounded. Torti understood immediately that the risk of water damage was real and urgent and he promptly went into action. Our construction heroes hammered away, closing off all the holes even more tightly, making them storm-proof. They covered all the beams with plastic. There was nothing more they could do for now.

But a couple of days later we noticed that the water damage had spread to the north wall of the same bedroom, which we had decorated with a lot of pain and effort. The lovely old-fashioned, rustic paint effect *di una volta* was covered in wet patches which would undoubtedly become mouldy and turn black with time. The whole wall would need to be re-plastered, but worse: it would have to be stripped bare. The thought didn't make us happy. Damn Torti with his promises that the windows would be put in within a week, no, within three weeks, no, within eight weeks, no, in how many weeks actually? This time we decided to raise the issue with our architect: maybe he could put pressure on Torti to do more to avoid water damage and especially to make some progress with fitting the windows.

Once Torti and Cassani both arrived on the site, the source of the trouble was soon identified: the guttering. It may not have caused the night-time banging, but it was the cause of the water damage. The down pipe was detached from the guttering on the roof, which meant that the water was streaming through the outlet straight onto the side of the house in one elegant waterfall. Water that had been collected from one side of the roof, a surface of more than fifty square metres. For hours on end. The bedroom wall on the top floor was *bagnato*, drenched, and all that water had seeped down on to our bedroom wall. Long live gravity. All we had to do was to disconnect the guttering outlet from the wall and the problem was solved. Three seconds of work after days of torrential rain. We were not happy. Torti was not happy either, as we would find out in the next few days. He wasn't happy that we had gone behind his back and got the architect to check on him.

Il laboratorio

A frail little man shuffled out from behind the partition screen. He had watery eyes and spoke softly as if he was out of breath. Even his heavy glasses with their thick lenses seemed to add to his fragility as they kept on sliding down his bulky nose. The *laboratorio*'s boss wasn't here at the moment, he said quietly: he was out on a job somewhere. The reception area of this workshop, christened the *laboratorio* by the *fratelli* Crosignani, was crammed with TV sets, covered in yellow post-it notes. They were sets that had been repaired and were awaiting collection by their owners. The old boss shuffled to the back room, muttering something about getting a piece of paper. Full of curiosity, we peeked behind the screen and saw a storage room full of electronics, printing plates, soldering irons and screwdrivers. This really was an electronics laboratory!

The frail little man noted down our names and telephone number with a trembling hand in illegible handwriting. When the boss came back, he would let him know that we had been here, he sighed. All this time, he was looking so weak that we were worried he would drop dead in front of our eyes. When we got back into the car, we took another glimpse back to see if he was still standing. Yes, luckily he was still on his feet, wiping his blotchy nose with a big white crumpled handkerchief in the doorway. We could barely bring ourselves to leave him behind on his own.

We had gone to the *laboratorio* on the advice of the *fratelli* Crosignani, who were *the elettrodomestici* suppliers of the area. We were looking for an electrician *di fiducia* to replace the scoundrel from Scorzoletta who never came back to finish his job, despite all the promises to the contrary by his colleague and partner in crime, the plumber. We needed a real electrician to check over all the colourful cables and connections that were tumbling out of the wall sockets. We wanted everything to finally be *a norma*. Even the thermostats, which were bought 'especially for us' by the plumber, were already ready for replacement.

Later the same day, we received a phone call. The voice sounded familiar. Wasn't that the soldering-Crosignani of the two *fratelli*? But the person on the other end of the line was ringing us to talk about the electrician's job we had requested at the *laboratorio*. Oh well, we thought, maybe he would pass on our enquiry to the *laboratorio*, which was right around the corner from the *fratelli*'s shop. The next day, a van pulled up in the drive and two men got out. They headed straight upstairs, bringing their tool kits and other equipment. The smaller and sturdier of the two informed us that they had come to look at the electrics. That voice again! Identical to the voice of one of the *fratelli*, but he wasn't here. This man also looked a bit like him if you looked closely enough. When we told him about our confusion, he revealed that he was one of the *fratelli*'s cousins. His father, the frail old man whom we had met, was the brother of the *fratelli*'s father. Great, that was all clear then. It was remarkable how strong the influence of genes must be, that the cousins looked more like each other than the brothers. Were their fathers twins by any chance?

The electricians got to work, which turned out to be quite a spectacle. The taller of the two looked a bit unkempt; he was unshaven, with greasy hair and dishevelled clothes. He was holding a roll-up cigarette nonchalantly between his lips, and it was shedding bits of ash onto our beautiful tiled floor from time to time, but the man didn't even seem to notice. At the same time, he looked at you with friendly, twinkling eyes. Whilst they were working, they communicated with each other using a language of gibberish that only they understood. Was this some sort of secret code that they had developed in their years' of partnership? They seemed to be in complete harmony with each other. One would pull at some cables while the other stood over at the main fuse box, observing the effect. Mumble, mumble, a surprised look on one's face, a questioning look on the other's. Time to try something else. And so on... The analogy was striking. They looked exactly like the world famous handymen duo from the Czech animation series Pat & Mat. We were being kept wonderfully entertained as we watched the activities of our technicians *di fiducia* from a distance.

In the end, they managed to solve all the puzzles without getting themselves embroiled in one of Pat & Mat's typical disasters... We also received new thermostats that cost a fraction of the price of what the *idraulico* charged for his dodgy devices. These were, according to the cousins, the best of the best: high tech! The thermostat was lined with rows of small, colourful switches: one for every hour of the day. Now you could regulate the temperature hourly! The trip to the Moon wasn't in vain! *'A je to'* as Pat & Mat would put it: And that's it!

Il malumore

Torti had made himself unbearable on the building site for the last couple of days. He was shouting and grumbling even more than usual and his attitude was driving his work-mates round the bend. He kept calling us over (*"Nico! Stefano!"*) to discuss the smallest of decisions. Well, to discuss.... It was more of a case of us having to approve all Torti's decisions. It looked like Torti was trying to cover himself against future complaints. It was only when he started to go on about the boundary wall (the old wall had been in a much worse condition than he had thought at first and it would require a lot of extra work) and kept on coming out with *"Ditelo all'ingegnere, ditelo,"* that it slowly dawned on us what was bothering him: Torti felt unfairly treated because we had contacted Cassani about the water damage issue behind his back. 'Tell it to your architect', was his message. Torti was clearly a man who needed a manual, one that was only available in Italian.

 We ignored his advice: why should we disturb Cassani unnecessarily? Torti knew all along that the old boundary wall was in a bad state, didn't he? Stop whingeing and get on with your work, we thought to ourselves and retreated to our plastic-wrapped fortress. But that couldn't last long because Torti was soon on our case again with another problem. His face was like thunder. One of his labourers led us to the source of the new problem. Behind one of the openings that had been made for a supporting beam of a new balcony, they found a pipe: a waste pipe leading out of the boiler. They couldn't insert a beam in there. What now? Torti asked. We referred him bravely to the *ingegnere* (*Chiedilo all'ingegnere*, we thought meanly: ask the architect!). Torti rang him reluctantly. In order to route the pipe around the beam, they would need to completely open up the wall. Torti didn't intend to do that, and neither did we. Instead, we decided, together with Cassani, that the balcony would have to be ten centimetres narrower. Everyone was happy. Well, nearly everyone. Torti was looking for more ammunition to stoke his discontent.

The channel had been dug for the foundations of the new boundary wall had to be dug deeper, in his opinion. Mimmo, the Sicilian, had to do it, using the pickaxe (our pickaxe) because the excavator had failed to do it. Mimmo wasn't happy, he was not supposed to be undertaking heavy physical work because of his heart problems. But Torti couldn't care less and gave him heavy jobs that weren't good for his health. Mimmo ended up just pretending to work whenever possible, cutting corners and stopping completely as soon as Torti was out of sight. We had just seen him climbing into the digger barely a minute after Torti had left the site. These little conflicts at work provided plenty to discuss during the lunch break that Mimmo and Mariano (the Moroccan bricklayer) spent on our terrace. Festi and Marco always went home for lunch in Marco's Fiat Panda. It was a strange sight to watch the tall Romanian folding his lanky body into that tiny tin can. He had to bend his head to be able to look through the window.

Mimmo produced the same backpack every lunch time and unpacked the contents: a Primus stove, an aluminium saucepan and a box of fresh pasta. The pasta had been prepared for him by his wife and he would warm it up at his leisure in his little saucepan. Mariano always brought a cold vegetable dish, sometimes with cous cous. He made his lunch himself because his wife was still in Morocco. Once in a while we joined in with our own Dutch-style sandwiches which Mimmo would eye up with suspicion. Bread with topping, an unknown concept to Italians. And a mug of milk! "*Siete fuori*, you are crazy," he said looking at us good humouredly.

The conversation over lunch was easy-going and occasionally we received some interesting snippets of information. Like the time when Mimmo and Mariano told us that Festi and Marco were paid cash in hand, despite all that trouble we had gone through to get the DURC (declaration of good professional standards)! Mimmo was complaining that even though he was a skilled *muratore* (bricklayer), he was also given all sorts of mundane tasks and Torti also expected him to do heavy physical work. He had already had two heart attacks! We were shocked to hear this, but Mimmo's tanned face smiled at us nonchalantly as he was speaking. He didn't think it would do him any good to get all depressed about it. He coped by getting on with his life. Except for the heavy jobs. He wasn't OK with that. He had worked as a self-employed *muratore* for a couple of years, but he found it hard to make a living on his own so he joined Torti's gang. "We argue nearly every day because I just can't keep my mouth shut," he said. "And that's not good for my heart either. Sometimes we don't talk for days on end, until we make some sort of peace that feels more like the Cold War."

Mimmo (Mariano was the silent type who only added a word or two now and then) told us how he was trying to enlighten the witless Festi about the nature of our relationship. "Where are their wives?" asked Festi him once, sniggering. "Well," started Mimmo, trying to explain it, choosing his words carefully. You had to be careful that you didn't shock someone so simple. But his efforts were wasted. Two men having a relationship didn't fit into Festi's view of the world. He couldn't even imagine it, things like that didn't exist. Mimmo found it hilarious. The silly Albanian!

After they had finished their pasta, Mariano and Mimmo often took a siesta until Marco and Festi's return. Then they all carried on with their work *piano piano*. It didn't take long before we heard Festi shouting "*Ė lui, è lui*" followed by the sight of Torti's car driving up the hill. What would he find to moan about this time?

La terza età

"*È brutto*. It's ugly," said Torti, looking moodily at the tree that stood near the end of the only remaining bit of original boundary wall. Torti wanted to cut the tree down, that was clear. Because the tree was in the way. Hence Torti defined it as 'ugly'. I didn't agree with him. I didn't know what kind of tree it was, but I did know that it was covered in pretty blossom every springtime. Beautiful, large clusters of white flowers appeared on it which had a lovely, fresh spring-like smell, much like wisteria. And the blossom looked like wisteria blossom but this was a tree with a trunk, not a climber. This tree was a common sight along the roadside, it grew on uncultivated verges across the countryside. With the car windows down, you could drive through fragrant clouds of spring air heavy with the scent of these blossoms. Heavenly. But the specimen on our drive was doomed to go on Torti's orders. Mainly to avoid it becoming a long-standing issue, because after the weeks of construction work and discussions with Torti, I had no energy left for one of those. There were plenty of trees like this in our area and if I really wanted one, I could always plant one in the garden, ugly or not. Maybe I should plant one in the place where Torti's much admired oak tree is standing now. I could cut that down, I thought, planning my revenge.

But what kind of tree was this exactly? That question was still unanswered. My enormous English plant encyclopaedia which contains thousands of pictures of plants and trees, classified by time of flowering and colour, could not provide me with a definitive answer. Well, that wasn't going to be easy then. But I got lucky. I had recently enrolled at the UniTre Broni, the *Università per la Terza Età*: the University of the Third Age, catering for people of near retirement age and beyond. It seemed to me a good way to meet some more Italians, to practise my language skills and to become a bit more integrated. Unfortunately, the name of the institute didn't quite correspond to the reality. That fact that it wasn't a university at all, but more like a hobby club run by good-natured amateurs, was of course not a disaster, maybe even an advantage. My intention was not so much to learn but to meet nice Italians with the same interests as I had, in gardening, cooking, lace making, etc. But what the concept of *terza età* meant in practice was that most of my fellow students were over sixty five. At the first lecture I went to, I saw a group of elderly Italians shuffling into the stuffy old auditorium. It was obvious that this group was not here because they were particularly interested in the topic of the lecture, but because they attended every talk organised by UniTre. It was more a place that provided occupational therapy for the elderly than an educational institution for the over-fifties. I couldn't see myself making many friends here. We didn't seem to have enough in common.

The lecture was surprisingly informative. The speaker was a regular contributor, who came at the beginning of each 'academic term' to show his latest nature photos. They were pictures from the Po Plain near Broni, depicting birds, plants and landscapes. I thought this would be nothing new for the old people who had lived here all their lives, but I found the slide show fascinating. I learnt that the little groups of poplars, planted in straight rows here and there on the plain, were indeed meant for logging. My tree (by now 'ex-tree') also made an appearance. The speaker mentioned it as the *robinia*, or the locust tree that was introduced to the area in the 19th century. It's as strong as hardwood, excellent for making plant stakes with. In the old times, people used it to prop up the vines in the vineyards (nowadays they use concrete or aluminium posts). But it proliferated and now it's all over the countryside. That solved my little mystery! Now I could finally teach Torti something new.

The rest of my career at the UniTre was short-lived. I did another course about roses, called '*Le rose, che passione*' which was attended by the same group of 'third agers', and which was taught by a prickly middle-aged woman. She spent a lot of time laying down the rules, explaining which topics we were going to cover and which we weren't, and what kind of contributions she was expecting and mainly what type she wasn't. It was plain that the course leader had already had some previous experience with this group that was not entirely to their mutual satisfaction. "*È permalosa*," I heard someone say in a private conversation during one of the breaks. When I looked it up at home - because I was a bit nosy - I found that it meant 'touchy', according to the dictionary, that it was easy to tread on her toes. There wasn't much passion exhibited at our lessons, apart from a passion for gossiping. I had already noticed, that Italians use the words *passione* and *emozione* a lot more lavishly than us cool Northerners, which rather devalues their meaning. After attending a cookery course, with a little club of white-haired ladies who had known each other for years (and made me feel like an outsider), I decided that the UniTre wasn't the place for me. Maybe I am not *terza età* just yet, I concluded joyfully. But now I knew the name of the tree that used to stand on my drive before Torti hacked it down for aesthetic reasons.

Persiane articolate

It was night-time and I was lying in bed in the future holiday apartment. The storm was raging outside, yet again. I suddenly heard a noise from upstairs. What could that be? I couldn't sleep any more now that I had heard that din, so I decided to go and take a look. In the dark, I stumbled up the staircase that was still covered in building dust (would this chaos ever end?). My heart was beating in my throat: the fear of the bogeyman seemed to be embedded deep in my psyche. A beam of light, coming from the study, was reflected on the floor tiles of the landing. Light? There were no lamps on this floor, let alone electricity. I crept quietly towards the doorway and tried to establish the source of the light. I saw building material and rubble that the workmen had left behind earlier today. There were bits of plastic sheeting and piles of bricks here and there. I looked through the closed, south-facing windows and saw a bright sky. It's a full moon, I concluded. Nothing to be worried about. I felt immediate relief. Naked, I took a step forward into the doorway, where I got the fright of my life: there, standing in front of the opening that had been made for a window (not fitted yet) there was a black silhouette! He was lit from behind by the moonlight so I couldn't make out the expression on his face, but I knew immediately who it was: Torti! "Nooo!" I screamed.

I sat up in bed, drenched in sweat from the nightmare that had just awoken me. Was that damn Torti going to start haunting me even at night? I started agonising about how I would have to ask him (or Cassani) for the hundredth time to get on with fitting the windows on the top floor. The iron frame, delivered by the blacksmith, had been successfully fitted - after one failed attempt - to Cassani's satisfaction. The plastering was finished and the door frames were in place too. The blacksmith could now fit the kitchen window and the French doors as soon as he had finished making them. But when would the blacksmith come? Torti told us that he would first need to come and take some measurements. Why? We asked ourselves. The windows just needed to fit the frames that the blacksmith had prepared earlier himself. Didn't they? Yesterday he hadn't turned up, much to our ever increasing frustration, despite Torti's promises to the contrary. Why not? I had to make it clear to Torti how important this job was, without being fobbed off again by one of his smooth excuses. The blacksmith had to come out today. Not *domani*, but *subito*!

But either Torti's guardian angel had been at work, or Mrs Torti had a crystal ball (with a detailed instruction manual from Torti) because this morning, Torti had once again escaped my tirade. As soon as he arrived he left again with Cassani, in order to pick up six eggs from his house, as we were told later. We had also received six this morning: Torti's chickens had, according to him, *un sedere di oro*, golden bottoms. And whilst he was away, sure enough, the blacksmith turned up! That was perfect timing because now we had the chance to have a proper chat with him about the windows, without Torti sticking his stubborn nose into the conversation. He had come to measure things up, the blacksmith said, because he had received all the material. He would get to work immediately, he promised us. We asked hopefully whether that meant that the windows would be in by next week. It wasn't that simple! He was now going to make the window frames and order the glass. That would take a couple of days. Of course! Glass had to be ordered. Everyone knows that. We understood fully. Take your time, dear ironsmith. We were trying to understand why this procedure had to be so complicated but the blacksmith's explanations didn't help. It was clear, however, that Torti's schedule for doing things was not efficient. He could have saved a lot of time if he had ordered all the material first and had only made the holes in the walls once everything had been delivered. We asked the blacksmith if he could already order the material for the small kitchen window and for the French windows if he measured them up together with all the other window measurements. "Hey, what a good idea!" he answered. He would never have thought of it by himself.

Soon, Torti and Cassani had returned to our crime scene with their golden eggs. We wondered what they could have been talking about while collecting all the eggs. Was Torti sneakily trying to get the architect on his side? You never knew. For an outsider, it was hard to assess Italians' relationships to each other. Cassani started with an inspection of all the work that was going on at the site. The blacksmith was busy taking measurements of the window frames that he had made and the ones he would be making. The conversation came to the wrought iron *ringhiere*, the balcony railings, yet to be made. We stressed that we wanted to choose the design. If we weren't careful, Torti and Cassani would make a decision behind our backs. Just the other day I had pointed out an excessively elaborate fence in our neighbourhood, just to tease him. It was a *disegno esagerato*, an exaggerated design, but Torti took me seriously and was now afraid that I really wanted something complicated for our balconies. He immediately lodged his complaints about the costs with Cassani. Tee-hee! To drive him up the wall even more (as a little revenge for lying awake all night) I added that our railings were not going to be made of iron, like that 'commonplace' fence I had pointed out, but of gold! Torti didn't find this amusing, despite being the owner of 'gold-bottomed' chickens.

Finally, the conversation moved on to the shutters that would need to be put on the outside, in front of the balcony doors. In Italy, every window has an exterior shutter that gets firmly shut in the evenings. The Italians think it's crazy how we Dutch people leave our curtains open in the evenings. You can look right inside! But we think it's crazy that in Pavia, for example, at Christmas time, people in the *condominis* go to great lengths to decorate their balconies with pretty, festive fairy lights, which they can't enjoy themselves in the evenings. That's a perfect example of *fare bella figura*: the decorations are meant exclusively for the outside world and not for the people who live inside the house.

Our shutters, *persiane* in Italian, had to be *articolate*, in other words folding. It had to be possible to fold them, otherwise they could never be completely opened: the door opening was too wide compared to the width of the balcony. You would end up with two hooks on the outside wall that you would never be able to use. That would be a bit silly. Shutters with joints which allowed them to be folded into two seemed to us the best solution. But according to Torti they would be too heavy (or too expensive?). Luckily, our architect, Cassani, was of a different opinion. But unfortunately, the blacksmith, who would manufacture the shutters and the frames, had yet another opinion. His concerns were of a technical nature. In the end we came to an agreement, after having visited another building site, where the blacksmith showed us what he could and couldn't do.

The windowless phase was coming to an end, after nearly two months of waiting. It wasn't a day too soon. According to the weather forecast, the first snow would fall by the end of the next week.

I tartufi

"Barbera," answered Torti as quick as lightning, before we even had the chance to open our mouths. The waitress was enquiring whether we wanted *bonarda* or *barbera* wine with our *polenta al cinghiale*. Torti was having pasta, not *polenta,* and he never touched wine, as he told us many times before. The waitress was evidently not addressing him. Yet, he gave a direct answer without having asked us, because *polenta* with wild boar had to be enjoyed with a nice robust *barbera*. That's what went with it. *Ciak, ciak*, went through my mind. There he goes again with his theoretical knowledge. On the way to the restaurant he had pointed out some of the vineyards which, according to him, were producing wine of lesser quality. Wine that he had never tasted and would never drink. Oh, it doesn't matter, bring on that *barbera*. We accepted Torti's intervention without objections and kept our thoughts to ourselves.

We were sitting on a terrace in Varzi, a little medieval town in the hills of the Oltrepò Pavese. We had spent all morning traipsing round the farmers market at San Sebastiano al Curone. On the way home, we decided to have *pranzo* (lunch) in Varzi. Having lunch with Torti. Who would ever have thought that would ever happen! Sure enough, he took centre stage and entertained us with all kinds of anecdotes from his youth. Stories that must have taken place nearly 65 years ago. Telling us about how in the old days there were a lot more street parties in the neighbouring villages, with games, dancing and music. This strong sense of community was waning in the region, according to him. With not a little gusto, he told us the story of how he and his friends had stolen sweets at one of these parties, by extinguishing the gas lamps one by one with a catapult. In the sudden darkness that followed the rascals had struck and run away with pockets full of sweets, more than they could eat. Torti was the sniper; his friends the light-fingered thieves. "Those were the days," sighed Torti. "You don't come across parties like that anymore."

We had got into his car early that morning (not 'Dumpy' but his grey Punto) to accompany him to the farmers' market where he offered to show us around. After some hesitation, we agreed to come with him, we had no good reason to refuse Torti's invitation. Undoubtedly, he would have been offended, if we had refused to come, and we were relying on his goodwill as long as our building work was still going on. That meant a day out with Torti. Mrs Torti waved us goodbye with a concerned expression. She couldn't join us because she had to look after all the pets, and maybe her dear daughter's cacti were also in need of some special attention today. During the journey, the old man muttered in a broad Italian dialect about other road users and speed restrictions, but after about two hours' drive we managed to arrive safely at our destination.

On the embankments of the Curone river there were hundreds of market stalls where they were selling cheeses, sausages, mushrooms, and, more importantly, truffles. Thousands of food connoisseurs came from far and wide to admire the delicacies: a crowd united in their passion for food. Torti led the way, his beckoning hand-gesture working overtime. He knew exactly which stalls to visit, and we were not supposed to dally and admire the gastronomic masterpieces at other stalls. You had to be careful where you shopped, he warned us. Some scales were tampered with and sometimes you were given much poorer quality *salame* than what you were offered as a taster. You would only find out later, once you got home. Torti's world view, the danger of being deceived, lurked everywhere. But there was also a bit of bravado involved in the presumption that he knew exactly where you could buy the best products. A characteristic shared by many Italians. Special, more or less secret little addresses (*di fiducia*!), recipes, ingredients, every Italian has them, but they all have different ones. On the way to the farmers' market, Torti showed us a side road leading to an *agriturismo* where, according to him, you could get the best apples in the region. He always got his from there, he said. As sober Dutchmen, we thought it was a bit over the top to drive a couple of hours for a crate of apples, when there is a wide choice of perfectly good quality apples in every local supermarket. Italians and their secret addresses.

Once we had squeezed our way through the crowd past every market stall and had arrived back at the beginning, we wanted to go into one of the side streets to look around the rest of the village. But Torti called us back: "*Non c'è niente*! There is nothing to see there." He wanted us to go to another part of the market because there was a stall selling wood burning stoves. We were planning to buy one of these and had been stupid enough to tell Torti about it. Now he knew, of course, exactly which kind of wood burner would be the best one for us! He would go and have a little chat on our behalf with the dealer at the stall. Entirely against our wishes, Torti started up a conversation with the trader. We stood reluctantly outside the stall, waiting for him. Reading the signs on the stall, we realised that the dealer came from somewhere near Bologna. We weren't going to buy a wood burning stove that would need to be transported from such a long way away. We would go to see the Crosignani, we thought, then we would soon be sorted! Now we had this tiresome Torti sticking his nose into our business. Once he had had his say, he pushed a bunch of information leaflets and the trader's business card into our hands. If we contacted this company, Torti assured us, it would all work out well. When, weeks later, he spotted a different wood burner in our living room, he shook his head gloomily once more. A *stufa* from the Crosignani, oh dear, oh dear, that will be the source of endless trouble. It seems that the *fratelli*'s shop wasn't on his list of *di fiducia* addresses. But it was on ours!

Patata in testa

I had never seen Marco, the tall Romanian, run so fast. He was upstairs, smoking and hanging out with his mates, when from the office window they spotted Torti driving up the hill in his dumpy truck ("*È lui, è lui!*" shouted Festi suddenly). In order to avoid giving the impression that they were dawdling (which they were), Marco made a sprint with his long legs downstairs to his abandoned drill. He was just in time, drilling away as Torti pulled into our drive. Whilst Torti was parking, Marco 'temporarily' took a break to help the boss unload the truck. Marco: always ready to lend a hand, *bravo*! The rest of the day Torti was on watch and everyone busied themselves with making the holes for the supporting poles for the panoramic terrace. We sat in the kitchen, which was the only habitable room in the house, and where all the noise seemed to be focused. The noise droned into our ears and vibrated through our bones for hours on end. But we could take anything, because at the end of the previous day the last windows had finally gone in!

"*Si chiude prima dell'inverno!* Please close them up before winter arrives!" I exclaimed at the time when Torti was about to break through the walls to make those damned holes. He laughed at me then, the old man, he thought I was a *buffone*, a joker. But the eight to ten days that he had predicted were to turn into about forty. Our patience was tested to its limits. Planning was obviously not Torti's *forte*. That was how we had ended up with six bare supporting beams sticking out of the west side of our villa for some time now: three per balcony. The delay was caused by the wait for the delivery of the marble plates which would form the floor of the balconies. Although, in the first week, Torti explained to us proudly the clever logistics behind getting the *marmi* in place ("*Viene la gru,*" he exclaimed every time and it took us some time to understand that he was referring to a crane), he had ordered the marble plates too late. Soon he would have to stop working on the terrace because this would hinder the fitting of the marble balcony floors. Torti's solution to this kind of *stasi,* stalling, was simple: he would suggest making a start on demolishing the future guest apartment too, even though this was our only habitable space. Had we refused, Torti would be stomping his feet and threatening to leave his crew at home. Until now, with Cassani's help, we had succeeded in preventing Torti from turning our living quarters into a disaster zone. We kept on coming up with new jobs for him. But would this trick work now? Let's hope the *marmi* would arrive soon.

Torti's workmen were regularly driven round the bend by his unpredictable actions. Invariably, Torti would announce the work schedule for the following day at the end of each day. But we had soon learnt to expect exactly the opposite: the work that had been announced would definitely not be done; instead, Torti would decide *all'improvviso,* on a whim, what he was going to do in the morning. "*Ha una patata in testa,*" Mariano would say on these occasions. "He has potatoes for brains." Our clever and able Moroccan often saw the solutions to certain problems well before Torti, but he was condemned to endure having to watch Torti quietly, whilst he - without taking any heed from his bricklayer - struggled on for another half an hour. He kept on fretting with his quivering tape measure: measuring, and measuring again, no that's not right, measuring again, from the beginning. Oh well, it's true, he was getting on a bit. Maybe at 74 he should be enjoying his pension by now. But according to Mrs Torti that was not going to happen any time soon. Her husband couldn't sit still for long. *"Vuole solo lavorare, lavorare, lavorare."* She said this with a tired but resigned sigh. The dogs, the chickens and the cacti would remain her responsibility for the foreseeable future.

Luckily, we could now start with the cleaning, scrubbing and decorating upstairs so that we could get our own apartment habitable again. The first part of the building project was over. We were hoping that the rest of the construction work would carry on without many delays but the signs were not promising: they were still forecasting snow, it would fall within the next few days. Torti had been muttering for the last couple of days about our architect, who should have let him start the work earlier, without that cursed DURC. Then he would have finished the terrace by now. The fact that the slow progress was also thanks to his chaotic planning was something he had never considered. *Patata in testa?*

Il meteo

At the end of one of the many endless talk shows on *RaiUno*, the main public TV channel in Italy, suddenly there was a soldier on the screen. Was this a military coup? We could barely suppress a politically incorrect sigh of relief. But no, of course not, this country is not a complete banana republic, after all, there is still law and order despite the large numbers of irresponsible politicians. In the short time that we had lived here, the 84-year-old, ex-communist president Giorgio Napolitano had become our hero. In spite of the turmoil that's created around the clock by Italian politicians (who lack for nothing in their creativity and shamelessness), Napolitano keeps the government's ship on a steady course. He resolutely dismisses unconstitutional government proposals consistently. Politely at first, but he follows up with a firm disciplinary rebuke if he has to. A coup was thus unnecessary because *meno male che Giorgio c'è*, we still have Giorgio, thank God!

The soldier on *RaiUno* started to rattle in long, full Italian sentences. "*Guardiamo adesso le previsioni relative alla situazione meteorologica e le immagini che ci raggiungono del satellite.*" He was trained to deliver the weather forecast within a couple of minutes because it was followed by the news, which had to start on time even in Italy (except in special circumstances like natural catastrophes and coups). Our weather presenter of the day was *colonello* Guido Guidi, one of our favourites, even if only because this gave one the opportunity to say: "*Guida Guido Guidi*. Guido Guidi presents," which just rolled off the tongue. It reminded me of *veni, vidi, vici*

(Guido came, saw and conquered) or of an Italian version of a tongue-twister like: 'she sells seashells by the seashore'. Furthermore, Guido was a classic example of a handsome man, one that every *mamma* wishes to have as a son-in-law. And as a *colonello*, he wore a uniform, always a good finishing touch. There are several other colonels and *tenenti*, lieutenants, who take turns to present the weather forecast. The main feature they have in common is that they all make a very trustworthy impression. Why? In Italy, *il meteo* is traditionally provided by *Aeronautica*, Air Force employees, which puts weather forecasting under the control of the Ministry of Defence: '*Le previsioni meteorologiche sono a cura dell'Aeronautica militare*' it says on the web page of the meteorology institute. So there was Guido today, in full military uniform, in front of some satellite images of Italy, rattling on about the disturbances and cloud formations visible in the pictures. After several weeks of practice, we became proficient in following Guido's presentation because it always followed the same structure.

But what was our gallant Guido announcing now? *"La prima neve dell'anno sta arrivando, anche a quote basse in pianura."* The first snow of the year was imminent, even in the lowlands and in the Po plain. Oh no, we were thinking, there goes our schedule. Or to be more precise: our delays. Torti was still working outdoors and was probably not going to be able to carry on with the terrace and the balconies. Maybe he could already make a start with opening a hole in the wall, downstairs in the live-in kitchen, providing the *padroni* with some fresh air and a lovely view of the snow-covered vineyards. The latter *in omaggio*, free of charge, of course. It was not that funny though. Imagine if the snow stayed through the whole winter! We could barely think about it and were hoping that our Guido had made a mistake this time. We would forgive him readily.

But we had no such luck. The next day we were greeted by a thin layer of snow. It was only a thin layer, but enough to give Torti and his companions a reason to skive off. Even though yesterday the first marble plate had finally been inserted into its place. They used the famous *gru*, which - instead of the spectacular crane we had expected - turned out to be a small rickety crane that had been sitting in the back of Torti's 'Dumpy' all this time. Inserting the hefty marble plate was quite a task and it required a concerted effort. Our construction workers were cutting into the slots in the exterior wall up until the last moment to make them a perfect fit for the plate. Torti was continuously barking instructions and warnings at them, which didn't really help the proceedings. But in the end, the balcony floor was in place. That is to say, half of it. Because the floor was made of two separate blocks which had to be bonded with an adhesive. It looked rather strange, half a balcony, without railings. Our blacksmith had – of course – to take some measurements again before he would make any railings. Luckily he already had the materials he needed in his workshop and we had already visited him, accompanied by Torti, to look at possible designs. Our crazy dream of having intricately decorated balustrades, embellished with vine leaves, was quickly scrapped because it was just too expensive. Instead, we went for a simple yet elegant standard design whilst trying to fend off as much of Torti's advice as we could.

Fortunately, the snow melted in the next couple of days. Torti could carry on with the building work. After some precarious manoeuvring with the *gru*, under Torti's frenetic instructions, our balconies finally had floors! The next big task was glueing the two halves of the floors together. For this to work, the glue had to be heated up. Did we have a hair dryer, asked Torti. "No," I answered, pointing at my bald head, "I haven't used one of those for years." We found a little electric heater left behind by one of the previous owners, but it was harder to manoeuvre than a hair dryer, especially on a balcony without railings! "If we are not careful, we will end up like Icarus," joked Torti, who was suddenly in a very jolly mood, now that he had completed his masterpiece. Thankfully, all the work was done, without any mishaps. After all, a DURC wouldn't be of much help if you ended up in plaster from top to toe.

At the end of the day, when all the workmen had gone home, we stepped through the beautiful wide French doors right into the panoramic painting of the Oltrepò. At this time of year, shrouded in autumn colours, the countryside looked like an impressionist painting with swirling clouds that would make a man like Guido Guidi jealous.

Il salvavita

The countryside looked like a scene from a fairy tale. The rolling hills and vineyards were covered by a soft white blanket of snow 16 inches deep. The fresh snow absorbed every little noise, creating a breathtaking silence. You could walk across the vineyards again without the sticky mud pulling your shoes off. Saar was leaping across the snow like a *capriolo*, a deer, otherwise she would have sunk into it. Once in a while, she took a bite of the snow: delicious fresh water! The virgin snow lay undisturbed apart from a couple of animal tracks here and there, which could have been made by a fox, a wild boar, or maybe a hare? We weren't experienced enough yet in tracking.

Indoors it was less pleasant because in the morning we had a power cut. Power cuts were common during storms and at night, but they never lasted longer than a couple of minutes. This time, however, we were without electricity for hours. Snow seemed to be a big obstacle for electricity provision. Electricity was still carried primarily via cables, supported by pylons high above the ground. No electricity meant no heating, and the live-in kitchen was slowly becoming colder and colder. After about an hour and a half, we saw our neighbour Piero shovelling the pristine snow from the pavement (the council's snowplough hadn't been along yet). We stuck our head through the kitchen window to ask him if he also had a power cut. We concluded from his incomprehensible muttering (he didn't half slur his words) that their house was also without electricity. Nothing to do but wait.

By now we had a beautiful, hyper-modern *stufa*, a fully automated wood pellet stove, delivered by the *fratelli* Crosignani. It was a colossus that weighed over a hundred kilos and was dragged upstairs by the *fratelli* with Festi's powerful assistance. It was a sight for sore eyes as it stood there in all its shining glory in our living room. It could make it lovely and warm in here... if it hadn't needed electricity. This was an unforeseen disadvantage to this modern take on the old-fashioned wood burner that we hadn't considered before. And we had gone through so much trouble to get Torti to install the flue pipe, which involved making a round hole in an internal wall. Torti kept on promising to do it. *"Lo facciamo, lo facciamo,"* he said but left every time without having made the hole. *"Facciamo due buchi,"* he even said once. *"Un'andata e un ritorno.* We will make two holes. One leading out, and one leading in." That was a builder's joke, but we couldn't really appreciate it because for the present there was no hole at all. Another time, Torti said that he would make the hole immediately if I fetched the *subito*, the metal ring which would be the finish on the hole. *"Subito,"* he commanded, and I complied, hoping that this would mean that he would finally finish the job. When I returned from my quick errand, *andata e ritorno* to the *fratelli*, Torti was already gone. I was furious.

Now the stove was installed, with flue pipe and all, but we had no electricity. We decided to shovel snow to get warm. We had to clear our own driveway, the council wouldn't do that for us. All the wine farmers were (very handily) using their tractors for this job, which they would normally use for working in the vineyards. We had no tractor, so we had to rely on manual labour. Fortunately, we did have a couple of snow shovels. Once the snow plough had cleared the roads, we would still be able to get out in the car. Maybe to a bar to warm up in, or to the supermarket to buy some provisions in case we got snowed in for some time. You never knew! Silly that we hadn't bought snow tyres yet because now we needed them and they were required by law. Actually, where were the snow chains we had brought over from the Netherlands? Probably somewhere in the impenetrable chaos in the cellar. Impossible to locate.

On our afternoon walk with Saar, another way of keeping warm, we got into a conversation with our neighbour, Antonio's Romanian workman. The poor man, we were thinking, he must be going stiff with cold in his little cabin in Antonio's big cold shed. But when we asked him if he was cold, he replied "No, the wood burner is on, come and take a look" and he led us to the little house which stood next to Antonio's. Inside, a lovely warm fire was crackling in the kitchen, filling the room with warmth. When we told him that we had a modern stove that was now out of action, the Romanian gave a chuckle. Sometimes progression results in regression.

Suddenly we noticed that the bare light bulb above the kitchen table was on. How was it possible that they had electricity? Didn't they have a power cut? "Yes we did," said the Romanian. "This morning. But it was over in half an hour." We were astonished to hear this and went home quickly to find out if the *luce*, electricity, was restored in our house too. There was still no electricity. But wait a minute! Yes indeed, after a bit of hunting, we found the culprit in the cellar. The first peak in the voltage had made the main fuse, *il salvavita* (literally the life-saver) blow. All we had to do was to flick the switch back on and we had electricity again. We had sat in the cold the whole day for nothing. Wonder what that silly Piero was muttering to us this morning?

Schola Regina Pacis

I felt a poke in my back. It was Torti. He was sitting behind me in church and was now clapping at the end of the parish choir's concert. Sitting? No: standing. Because everyone stood up to applaud. I didn't because I didn't want to confirm people's preconceptions. The Dutch are famous for rising as one to deliver standing ovations after each performance. They are rarely known to do this abroad and to demonstrate how integrated I already was in Italian society, I decided to remain seated. Torti was apparently of a different opinion: 'Get up, you!' said his poke in my back. OK, I will get up, o great and wonderful Torti, I thought to myself. As long as you make sure that your work on our house will also be worth a standing ovation.

The applause was meant for Schola Regina Pacis, the choir that Nico had recently joined. How did that happen? On one of our walks, he started up a conversation with the married couple who lived in a beautiful house at the top of one of the hills nearby. When Nico walked past with Saar, their dogs started to make so much noise that the woman came out of the house to see what was going on. They got talking and one thing led to another. Before long, the future chorister was having a cup of iced tea in Salvatore and Fernanda's kitchen. As he talked about his interests in broken Italian it transpired that both Salvatore and Fernanda loved singing. They were members of the parish choir which rehearsed in Don Luigi's church, only five minutes' drive from our house! Moreover, the Schola, like many other choirs, was in need of more tenors. They agreed that Nico could come and watch one of their rehearsals to see if he liked the choir and their repertoire. That same afternoon, Nico received an e-mail from Enrico Vercesi, the choir master, who also sent him some music scores to look at. Nico decided to join the choir without further hesitation. The choir usually sang the customary church service repertoire, but once in a while, they gave real concerts where they performed songs of their own choosing.

By the way, the standing ovation they received that night was by the way well-earned, because they had really blown the audience away with their performance. The fact that they managed to reach such high standards, despite the narrow pool of talent that a small town like Scorzoletta had to offer, deserved *complimenti*. Don Luigi's richly decorated church provided the perfect acoustic background for the production of these wonderful sounds. The charming, melodious music was composed by the choir master himself, Enrico Vercesi. It was more than worthy of being listened to. It was a shame that the concert was not well attended: there were only about thirty grey-haired men and women. It was several times more than the number of people at the previous year's performance when only six people turned up in total.

Since the concert took place in Torti's hamlet, he and his wife had put in an appearance. Torti was in good humour again despite not being able to make progress with the building work recently because of bad weather: according to Guido Guidi, whose weather reports we never missed, levels of precipitation had broken all records. This resulted in the loss of about three weeks of possible work days. But that couldn't dampen Torti's spirits because, at our last meeting with Cassani, Torti had received a substantial amount of further work. These were jobs which were not foreseen at the planning stages of the building project. He would have to build a car parking place, with a little wall at one end to prevent landslide off the hill (Torti loved building walls). He also had to assemble our Dutch swimming pool and create a sunbathing terrace as well as finishing the driveway. This meant that in the future he could write out another tidy little bill.

At the end of the concert, the choir members went to have lunch at the *bar-ristorante Scuropasso*, accompanied by Don Luigi. This establishment is within the parish boundaries (Don Luigi likes to give all the restaurants in his village equal attention) and it's a very popular lunchtime destination: every time we visited it around lunch time, we always found the car park full. But where were all the guests, we wondered, when we entered the place with the choir? We only saw a bar and a couple of little tables along the wall. But the waiter led us through a narrow passageway and suddenly we came to an enormous hall, lit by neon lights, where you could feed at least a hundred hungry people. This is a typical Italian phenomenon, which you can come across in any village: restaurants which are equipped for large amounts of customers (e.g. wedding parties) and where even an elephant (i.e. the spoiled Italian *bambini*) couldn't cause much damage. We were served a typical Italian lunch: a choice of starters, two kinds of pasta, a meat dish and dessert, accompanied by plenty of water and wine.

At a nearby table sat Leda, the master chef of the restaurant Bagarellum where we had dined before. After tasting the delicious food at her restaurant, we came up with a daring plan to ask her whether she would be prepared to give cooking lessons to our future guests. This was the moment for us to act. Would she want to do it?

And sure enough, Leda said that she had already thought of giving cookery courses and she would love to take this opportunity! We agreed that we would arrange a trial lesson with some of our family members when they came to visit. We were over the moon: we knew for sure this would become one of our top attractions! This initiative was definitely worth a standing ovation, one that we didn't need Torti to poke us in the back for.

La tela di Penelope

If we were expecting to find relative calm during the building work that followed the breaking, drilling and demolishing phase, then we were to be disappointed. We woke up early each morning to the arrival of the workmen and Torti's shouting. Even before we could wipe the sleep from our eyes, they were already drilling and cutting into concrete walls and floors. And to think that Torti had started the demolition work on the very first day of our *grande lavoro*! He must get great enjoyment from the demolishing part of a building project, maybe because this is the only part that doesn't require taking measurements. If so, he could really indulge himself working on our project. In spite of all the preparatory work, even more cutting and drilling were necessary, every time something was installed or assembled. The slots in the wall needed to be cut to size in order to install the marble floors. Mounting the railings involved drilling into the marble blocks and joining them to the walls. The shutters required hooks on the exterior walls for them to be fastened to when open.

But the main focus of Torti's biggest drilling and demolishing project for some time now had been the cellar, where flooding was a regular problem. One morning shortly after we moved in, we found an indoor pool inside the *cantina*. Our boxed up swimming pool was floating around in it. That was a strange sight: an outside pool that was floating in an indoor pool! It had rained heavily the previous night, and at first we thought that the rain must have been blown through the gaps in the badly fitted windows. But later we noticed that small puddles appeared even when it hadn't rained hard and that their usual location was the middle of the cellar. It turned out that ground water had built up under the house, and the water pressure pushed it up through the floor. How did this affect the stability of the house, we wondered. Would we end up like some of the house buyers on TV programmes like 'A Place in the Sun' who sometimes get embroiled in disastrous building projects ("How stupid can people get?"). Programmes which we used to watch from our armchairs with a guilty pleasure. Luckily our architect Cassani was a specialist in water irrigation and drainage problems, of which there were always plenty in the rice fields around Pavia and its neighbourhood. He advised us to dig channels, to install drainpipes and create some drainage pits.

Torti was busy for days on end, drilling into the walls around the cellar and into the driveway and the cellar floor to find the source of the groundwater and to find out where to position the drainage pipes that would lead the water to the main sewage system. He made a hole in the cellar floor which showed us the level of the groundwater: it came up nearly to the level of the floor. One day Torti drilled in the right spot: suddenly, water squirted out from underneath the house and ran down the driveway in a little stream. Once this spring had as good as dried up (a small stream stayed behind that never completely dried up), we saw that the floor in the cellar dried out. At the same time, our house hadn't been washed away and hadn't collapsed. It would all be all right in the end. And just to make sure of it, Torti started drilling and digging again, because now it was clear where the drainage pipes and pits would have to go.

The biggest hole that Torti was allowed to make for us in recent times (and we don't resent this, we don't want to deprive anyone of their pleasure) was in the floor of the stairwell. This new stairwell gave us access to our own floor and we didn't need to disturb our future guests when we left the house or when we arrived home. To install the staircase, a suitable hole had to be made in the floor. Torti had just the right CV to accomplish this task. Unfortunately, this part of the project also turned out to be a never-ending story. Torti had to restore and finish off some of the floor after he had broken through it so that the steel frame of the staircase could be installed properly. This would have required him to take measurements and to use his brain while there were so many tempting demolishing jobs to look forward to!

Now that we were well into our building project and there was no way back, it happened more and more often that the workmen didn't turn up in the morning. The first time this happened, we enjoyed the silence but our initial relief was soon followed by worries about the progress of the project. We were concerned it would take a lot longer than planned. At our last meeting, we forced a promise out of Torti: he vowed to finish all the work by the middle of March. How was he going to fulfil this promise if he didn't even turn up on days when the weather made work possible? There were already plenty of days when they couldn't do anything because of the rain or snow. We didn't get it. When it seemed like Torti was going to skive off again, we rang up Cassani, who would chase him up with growing reluctance. "Builders are a *razza maledetta*, a cursed race," our architect once allowed himself to say. Within about half an hour of one of these interventions, we saw Torti and his gang driving up from the valley. Once they arrived Torti gave nothing away. We often found out where they had spent the morning from quiet whispers behind his back with our Sicilian *di fiducia*, Mimmo. They were working on a job in the council churchyard, or for the Count of Vistarino, etc... Until they got rounded up again said Mimmo grinning. He knew exactly what was going on.

In the end, we got so frustrated at Torti's absences and the slow progress of the stairwell (even the blacksmith had finished the steel frame some time ago and it was ready to be installed) that we decided to look up 'never-ending story' in *il dizionario*, in the dictionary. It turned out to translate to *'la tela di Penelope'* (Italians know their classics): it was the embroidery Penelope was working on, and which she couldn't finish as long as Odysseus continued his quest on foreign shores. We wrote up this little proverb on an A4 sheet of paper and pinned it on the centre of the cellar door. It had no effect on Torti whatsoever. Being a thick-skinned *maledetto*, he must have come across this sort of situation before.

È rosso!

The temperature was positively freezing and we had already spent about three-quarters of an hour waiting in front of the *Radiotelevisione Italia* (RAI) building in Milan. "We", being us, two *padroni* and our friend and ex-landlord Giorgio with his girlfriend. We hadn't seen each other for ages (the building work had taken its toll on our social life) but Giorgio, who was going to watch the filming of a television show, took this opportunity to invite us to come with him. He knew that we appreciated cultural chat shows like *Che tempo che fa* ('Today's world') or: 'How times change'). This show, presented by Fabio Fazio, was pretty much the only Italian TV programme worth watching. The soaps and shows with leggy blondes (known as *veline* in popular language) that seemed to take up most of the broadcasting time on TV weren't meant for us. But Fazio's show only had interesting guests, well-known people from culture, science and politics. Even if you didn't know them at the start of the show, it was always worth getting to know them. Moreover Fazio, unlike other RAI presenters, seemed to take a genuine interest in his guests and he asked intelligent questions. Not surprisingly, Fazio's show was often at the centre of small political clashes because for some inexplicable reason 'the right' often considers cultural interest and intelligence as the characteristics of those leaning towards 'the left'. Berlusconi's political allies had tried several times in vain to get the show taken off the air. The popularity of the show and its presenter had saved it every time: over four million Italians would rather watch this show twice a week than the drivel on other TV channels, most of them under Berlusconi's control.

Che tempo che fa contained plenty of humour and satire: halfway through each show, comedian and actor Antonio Albanese would come on and perform one of his inimitable parody sketches. Our favourite caricature was that of a French wine expert. After much graceful swirling, rolling, sniffing, gargling and unintelligible French muttering, he always came to the same surprising conclusion that the red wine in his glass was, well ... red. We laughed at this many times together with Giorgio, at our old apartment in Via Moruzzi. One of our favourite private little jokes was sommelier Albanese's "*È rosso*".

And now, thanks to Giorgio's invitation, we were going to be present at the recording of one of the episodes of our favourite show! We registered by e-mail with Suzanna, who asked us to arrive at the entrance to the studio by 2:30p.m. at the latest. By now, it was quarter past 3 and the biting winter air had chilled the waiting group of fans to their bones. We stared longingly through the studio windows, at the spacious entrance lobby, which looked invitingly warm. But we weren't allowed to enter yet. Fortunately, we managed to thaw out temporarily in a sandwich bar, where we took refuge and had a bite to eat before rejoining the rest of the group, all stiff with cold. Even when the door opened we would probably need to wait a little longer for our turn since we weren't exactly at the front of the queue. Yet it seemed we were in luck because as soon as they opened the doors, there was no more time to waste: the show was about to be recorded! We all had to register our forenames and surnames and declare that we didn't object to appearing on the screen. Would there really be people who sign up to come to watch a TV show being recorded but who would themselves refuse to be on TV? We had no objections, quite the contrary, we wanted to be shown on the screen: the more often the better! There was a tumult of people scurrying around with forms, pens, coats and bags. The latter were not allowed in the studio and had to be put into lockers. We had read on the RAI website that this spectacular event could be attended every week. This was surprising, given the lack of organisation we witnessed. For us, this one time was going to be enough.

Seating the audience in the recording studio was just as chaotic, but we were lucky: even though we were among the last to enter, we were given front seats. Our faces would be guaranteed to be on the telly now! The long wait in the cold was soon forgotten. But just a minute: the TV usherette noticed that the bald, shiny head of tallest of the four of us was towering above the rest of the audience. He needed to be moved from the end of the row. We had to do a bit of mixing and matching. For reasons known only to them, Giorgio and his girlfriend did not want to be seen together in front of the whole nation (and accordingly, did not sign the declaration!). After regrouping, the girlfriend and Nico became a nice heterosexual couple while Giorgio and I became a *coppia di fatto* (the Italian version of a homosexual couple). We were ready for it. The show could begin. Now, where were the cameras? Were we on screen?

The first thing that struck us was that the studio was in reality quite a lot smaller than it seemed on TV. Fazio's mobile desk seemed to have shrunk to half its size! We were also amazed by how relaxed the host was in the minutes right before the start of the show. Everything was recorded in one go, without any hitches. This time, the guests were the singer Noa, a well-known TV presenter and an 82-year-old actor. The actor was incredibly camp, quite openly gay. He kept on flirting with the audience (with Giorgio and me?) as if to say 'aren't I cute?': "Once, in the 50s a TV station had asked me if I was a socialist by any chance. If I was they would never employ me, so I answered 'no' I wasn't. That wasn't a lie because, in reality, I was a communist!" As a well-trained actor, his timing was impeccable. He was an old charmer who could still command an audience's attention, who laughed and shrieked (Giorgio?). Still going strong.

At halftime, host Fazio announced Albanese, our favourite comedian. Fazio explained to the great disappointment of the numerous fans in the audience that it was Albanese's last contribution to the show. But to soften the blow, for one last time, Albanese would perform his most popular sketch ever: the sommelier. Today, of all days, when we were sitting just a couple of metres away him! He was hilarious as usual, rolling the glass in his palm, swirling the wine in a snobby fashion, gargling, tasting and finally announcing: "*È rosso!*"

L'elicottero

"Careful you don't put salt in it," said Torti winking merrily at *l'ingegnere* Cassani. Every time we had a meeting to discuss the *progetto*'s progress, he made the same joke, referring to the first meeting when we put salt instead of sugar on the table which ended up in (and ruined) Cassani's *espresso*. Torti liked to share a good joke (on his own terms of course) and he was up for a good little chit-chat. He was, as our neighbour Francesco called him, a *capo pigro*, a lazy boss: he liked to dish out instructions to his team but he wouldn't make his own hands dirty. He would rather chat about delicious meals and how to prepare them. We already knew by heart the recipe for his famous *gnocchi di patate, ciak ciak*.

This morning, Torti was full of news about one of his latest projects: a restoration job on Count Vistarino's castle. He would need to start it the coming week. "Oh, but well, we are not quite finished here yet, are we," he said beaming; looking at Cassani and us with his twinkling eyes. He didn't seem perturbed by how the Count would react if he didn't turn up on the first day. We didn't care either as long as he would carry on with the work on our site. Whilst Cassani accepted our coffee and put sugar in it (after a quick check to see if it wasn't salt), Torti decidedly refused our offer to pour him an Italian brew. Has he also started a coffee-abstinence? Indeed, he didn't drink coffee, only *orzo*, he said. We have already come across *orzo*. It was a barley brew that, according to us, was only meant for the hardiest of people. Naturally, we had no *orzo* at home, so Torti had to make do without.

The four of us were sitting around the kitchen table, and once we had established the new course of action, we were trying to figure out how much all the extra jobs would cost altogether. Torti's temporary *preventivo*, quotation, led us to think that he planned to open a brand new hole, but this time, (instead of inside the house) in our bank account. It was clear that we had to put a limit on how much he could spend and find some areas where we could make some cuts. We could say goodbye to our plans for a natural stone *pavimento* for the driveway, which would have earned Torti a neat sum of ten thousand euros. What should we do with the driveway then? Cassani had an idea. How would we like a concrete driveway which could be finished off with a nice lick of paint applied by an *elicottero*? "Like that shade of green," Torti butted in, pointing at our acid green armchair. Were they being serious? We imagined helicopters whirling in the air above the vineyards on sunny summer days, but of course we had got the wrong end of the stick. Nothing like that, Cassani said. An *elicottero* is one of those machines that cleaners use to polish the floors of offices and large halls with. There was a machine of a similar design which spread paint on concrete floors and it was quite cheap to hire. This was our chance to get that violet paint out, we thought to ourselves. Maybe it was a job for our old friend Olita, the estate agent!

Another area where we could cut costs was the exterior decoration of the house. After some tapping and rapping here and there, Cassani and Torti unequivocally declared that the whole of the external plaster had to come off, and not just a part of it as we had earlier hoped. That was going to be a big bite out of our budget! Once the rendering was done, the plaster had to be painted, or, to be more precise, sealed, primed and painted with two layers of masonry paint. I estimated the overall surface area of all the external walls to be nearly four hundred square metres: a very labour-intensive, and thus expensive, job. Nico took his chance to suggest that we could do the painting (us? Really? I thought, with not a little trepidation). To my great surprise, Torti immediately agreed. I bet he relished the idea of watching me clamber around on the scaffolding. Suddenly, my mind was flooded with flashbacks to the traumas I suffered at secondary school on the wall bars. Once I fell off one of these torture devices and broke a wrist.

The bottom layer of the external walls would also have to be waterproofed against splashes. You often saw houses which were tiled along the base and we quite liked the look, but it would cost a lot more than Torti's *strollato* idea. *Strollato* was cement mixed with gravel which was plastered onto the walls in several thick layers. It wasn't particularly pretty (the examples we had seen in our neighbourhood reminded us of layers of vomit smeared along the bottom of the houses) but maybe a lick of paint in a cheerful colour could redeem it.

Even though Cassani summed up all the points we had agreed on several times, we knew we couldn't sit back and relax because Torti would just go ahead and do whatever he pleased. There were times when he threatened to do things that we hadn't yet asked from him, and other times he forgot to complete some work that we had explicitly asked him to do. When Mimmo, for example, started on the preparation work for the little pillars which were designed to decorate the terrace, he drew three circles on the terrace floor. "Weren't there going to be five?" we said to each other. We dived into the contract and there it stood in black and white: '*quantità*: 5'. Hanging over the new balcony railings, we asked Mimmo whether it was still the intention to still add the other two pillars according to the original plan. With a small ironic smile he nodded in the direction of his boss. After a bit of grumbling ("Yesterday you said it was three!") he grudgingly did as he was supposed to do. The only option was to watch him closely because he paid attention to everything except our agreements.

Would we succeed in keeping Torti on the straight and narrow and manage to rescue our budget? We were entering the exciting last phase of our *grande lavoro*. It was time to choose some nice colours for our home because it wouldn't be long before we found ourselves ten metres up on top of some scaffolding. Brrr.

La scala

The work on the damned staircase progressed at a painfully slow rate. Sometimes it felt as if Torti crept back stealthily in the black of night to undo all the day's work, just like in the story of Odysseus, where every night Penelope keeps on undoing her weaving. Did Torti have something against finishing this bit of the *progetto*? Creating the stairwell took an inconceivably long time. They kept on tinkering with it, making adjustments here and there without any visible effect. They inserted supporting beams, built little additional walls, created brick borders, did some plastering, etc... The work was delayed by long breaks at regular intervals because even if the previous day's work had not been completed, Torti would start every new day with new plans. It was driving us crazy just to watch it. We talked to Cassani repeatedly and aired our feelings about the situation, but he had as little power over the old man as we did. Had Torti's potato brain turned into mash?

The blacksmith had delivered some very solid-looking bits for the staircase quite some time ago, but they now lay abandoned, rusting among the accumulating rubble. Every time we saw them lying on the driveway we wondered in exasperation why Torti still hadn't fitted them. "Finish something for once, Torti, for goodness' sake!" In the end, after another week or so, our patience was rewarded. Torti summoned the blacksmith and with concerted effort and a bit of extra drilling and slashing, the staircase was finally in place. As far as we could judge by the sound, that is, because we didn't dare look. It was bad enough hearing the banging and shouting from a safe distance. After a couple of hours the commotion suddenly died down and everything went suspiciously quiet. Once in a while we saw the sweaty figure of the blacksmith as he went over to his van to fetch some tools, only to disappear again in the basement for long periods of time. Torti's labourers carried on with their previous tasks. There was nothing the matter, was there?

The staircase project – as we were calling it by now – had taken long and thorough preparations and we hoped it would be finished by the end of the day, finally providing us with our much needed private entrance to the house. Torti's workmen would not need to trample across the guest apartment with their dirty shoes caked in mud and concrete anymore. This meant that we could begin cleaning and smartening up that part of the house. Especially since they succeeded in fitting an entirely new (and fully operational) window and a new door in the kitchen within the space of a week! That just proved that it *was* possible after all! The ground floor was habitable now, in principle, and we could carry on with painting and decorating as soon as the workmen didn't need to use it as a through-way. This gave us a psychological boost because we were nearly at the end of our tether. Torti and his men were ever present, day in and day out: it was as if we were married to them, as if we had chosen to spend the rest of our lives with them. It was about time for a quickie divorce.

As the end of the day was approaching, we couldn't contain our curiosity any longer. We decided to check out our beautiful new staircase. The first scene that met our eyes was a hallway covered in dust and rubble. The next was an empty stairwell. Down in the basement, in the little hallway built with such finesse by Torti and his men, sat the blacksmith – his face flushed with effort – bending over a heap of metal bits and pieces which we recognised as the (now dismantled) solid metal staircase frame. As it turned out, in spite of all the planning and measuring, the frame did not fit. Even additional drilling into the basement floor (accompanied by Festi's loud cursing) couldn't create sufficient extra space to accommodate the frame. This resulted in a slight shift in the angle of the frame which meant that iron brackets which were supposed to hold the steps were askew. The construction wasn't level, it wasn't *in bollo*, as Torti must have undoubtedly pointed out to the blacksmith as he was supervising proceedings, spirit level in hand. The blacksmith spent hours removing every bracket from the frame that he had so carefully soldered together. We decided not to disturb him in this important work and sauntered away, disillusioned. Just before dusk, we heard the engine of a van starting up. It was the blacksmith on his way home. From the window of our apartment we saw his car climbing up the hill. In his trailer there was the sorry pile of the bits of our staircase frame, looking like the carcass of a prehistoric beast unearthed in the basement.

A couple of days later the staircase frame was finally in, containing newly soldered brackets in all the right places which were now *in bollo*. It was a shame that the *falegname*, the carpenter hadn't finished the wooden stairs (did we order them too late?) in time. It was also a shame that when the stairs were finally delivered, the screw holes in the wooden planks did not line up with the brackets on the iron frame.

La cipollata

Some people were waving at me from the terrace of the large yellow house along the little local road. Among ourselves we called this house 'the Plantation House' because of its symmetrical design; it had two wings (with their own little turrets) built on either side of a central entrance hall. We often wondered what this house was used for and who lived there. It was far too large to house just one family. Could it be some sort of residential care home for the elderly, the kind you came across a lot in Italy? In any case, it seemed that the people who lived there owned several dogs, and they were the small, yappy kind. When we walked our dog Saar along the long hedge bordering the property, the air would reverberate with barking, growling and breathless panting. Saar never reacted, just sauntered on undisturbed, but the snappy little beasts kept on running around like headless chickens. 'Panting behind the hedge,' I used to call it. I already saw a neighbour watching us gloomily. "*Mi rompono i coglioni,*" he said. "This barking is driving me crazy. Dogs are supposed to bark at night, not during the day!"

    Exactly halfway along the hedge there was a gap for a little gate, through which I spotted the terrace where the people sat waving at me. They had heard me coming long before I got there because pretty much everyone who lives in this neighbourhood owns a guard dog which starts barking as soon as they catch scent of Saar. They knew it was time for the *passeggiata di quel signore con la cagnolina con la palla in bocca* because Saar always brought her tennis ball along to our daily afternoon walk. The walk always led us along 'the Plantation House', then past a wine farmer who had two big barking *spinoni* also hidden behind a hedge. Their big hairy heads were a comical sight as they peeped through a gap in the vegetation. The walk always ended in a little meadow with cows in it, owned by the only cattle breeder in the area. Once we had finished our ball game we turned around and went back along the same route, past the same houses, guarded by the same dogs.

But now, as we were walking past 'the Plantation House' again, the people on the terrace were waving so enthusiastically at me that I stopped in front of the little gate for a moment to find out who these people were who were giving Saar and me such a friendly greeting. There were about five of them sitting around a table laid with wine and sausages. They seemed to be nodding in my direction, beckoning me: Come and join us! The food looked very tempting and I was about to go in, but Saar refused to go through the gate. Parading provocatively along the hedge while the yappy dogs were safely tucked away was all right by Saar, but she wouldn't enter their territory. No, that was too much to ask. When the other dog owners saw us hesitating uncertainly at the gate, they sent their own four-legged friends inside the house, which left Saar in charge of the garden. Saar had no one to compete with; she was first in line in front of the plate of sausages, as was I.

The company around the table was made up of the cattle breeder (whose cattle provided the sausages) and the three occupants of the house: a married couple and a thin man with a paralysed (polio?) arm. We had a friendly chat while enjoying the delicious sausages. We weren't talking about anything specific but the atmosphere was warm and jovial. The presumable source of this relaxed atmosphere could be seen, shining and prominent, on one of the little tables: a large bottle of *spumante* that was all but finished. I expressed my surprise that a large house like this one had only three people living in it. "Our daughter was going to come and live in the other wing," answered the owner of the house, "but after it was built, she changed her mind and didn't want to live *in collina*, in the hills, any more." Her side of the house wasn't quite finished yet so if I had some money left over we could create a nice little extension to our B&B in there, suggested the *paterfamilias*. "We've run out of money!" I answered, quick as a flash. "*Sono finiti i soldi*", using the expression I learned at the post office in Pavia. There was a gigantic basement running under the whole length of the building, so if I changed my mind..., he carried on. There was a gym without gym equipment (but if I had money to invest...) and there was even an indoor swimming pool... without water. But we already had a swimming pool without water. It was time to go. "See you tomorrow!" I called.

A couple of days later, we received a proper invitation from the Plantation-dwellers to come and celebrate a *cipollata*, an onion party, which they organised each year. Everyone brought their own dish, which had to contain a special type of sweet red onion as one of the ingredients. This kind of red onion was only available at a certain time of year. When the day came, we saw that as well as the occupants of the house and the cattle farmer (who had attended the previous little binge), there were other people too, amongst others our neighbour from across the road. He was a Macedonian who had worked as a guest worker and grape picker for a couple of years and then decided to settle down in Italy and become a wine farmer. We had already come across him a couple of times whilst walking through the vineyards, where he was always busy pruning and tying up vine. *"Come mai da queste parti?"* he asked in surprise on those occasions, because he had never encountered people walking through vineyards for fun, covering large distances from home. Once he even went and brought us two bottles of his wine on his scooter. We always thought of him as a nice, spontaneous man, but later we discovered that we had caused him a lot of frustration for quite some time. All the onion experts had outdone themselves; there was plenty of delicious food and a relaxed, friendly atmosphere.

Our presence quickly prompted a discussion about Francesco, our ill-natured neighbour The verdict was unanimous: Francesco was an impossible, annoying man who picked a quarrel with everyone and was impossible to reason with. The Macedonian told us a little anecdote about how Francesco had once waited for him when he was driving his tractor to one of his fields. Francesco stood with his arms folded and his legs apart on the unpaved track between the vineyards. These paths were open access for public use but Francesco claimed that this particular path belonged to him because he owned the vineyards on either side of it. He was of the opinion that the Macedonian should just go the long way round. If he refused, Francesco would simply shoot at him next time. They had had many arguments on this subject before but this time Francesco was physically blocking the path and was making verbal threats. He said he wouldn't step aside. Our Macedonian was in no hurry and wasn't going to let himself be intimidated by Francesco. He stayed put in his tractor and waited patienly. After about fifteen minutes had passed, Francesco decided he had had enough and left.

We asked the Macedonian where this argument had taken place. He said it was on the path that led from the end of our fields into the vineyards and which you could only access from our land. In other words, the path that used to be accessible from our land until Francesco asked us to put up a border fence, which we did very compliantly. Francesco led us to believe that if we didn't stake our claim to our land then someone else would. That's the Italian way, he said. Luckily, the Macedonian didn't blame us: now he drove around the vineyards, using a stretch of the asphalt road that led past Francesco's house. Luckily Francesco hadn't shot him yet.

Storto

Torti was putting the last finishing touches to the house. Finally all the windows and doors were in and the basement stairwell *and* staircase were completely finished. There were only a couple of outstanding jobs left: the terrace floor and the driveway had to be concreted over and all the external walls had to be plastered. But before any plastering could take place, it was going to be necessary to remove all the old, loose plaster. In most places the damaged plaster came away easily but there were patches where it could only be removed with the help of a drill. This meant that every single day for a week, from early in the morning to late at night, the whole house shook with the sound of drilling.

As my luck would have it, just before all the drilling was about to start, I stepped into the shower cubicle one morning and as I turned my head to the right I heard a snapping sound, like the crack of a whip. Pain shot through my neck, up towards my skull and exploded inside my head. For a moment I was convinced that my brain would literally explode from the unbearable pain. The worst was soon over and I decided to step back inside the shower cubicle. No sooner did I do so than I felt a surge of nausea and dizziness and the headache returned with a vengeance. I was panic-stricken, I was convinced I was having a stroke and was about to kick the bucket ("He was in his prime," they would say). Or maybe I was going to be paralysed from the neck down and in the future all my writing would need to be done on a PC that responded to eye movement. I dried myself off with difficulty and got more or less dressed, which was hard because I couldn't bend down. Bending down caused the pain to return, shooting through my head violently. I stumbled into the living room 'with my head held high' and said to Nico in a panicky voice, "We need to visit the *Pronto Soccorso*, the first aid station!" He thought it would be more sensible to see if I felt a bit better after I sat down for some rest. I had to admit this was a good idea because I wasn't sure I could make the bumpy car journey across the Oltrepò in my present state.

I spent the rest of the day mainly in bed, dosed up on paracetamol and plagued by new worries about neck hernias. The next day wasn't too bad, after a reasonable night's sleep things seemed to settle down quite a bit. The pain had eased and I could move my neck more. But I still couldn't bend my head, which caused problems when climbing the staircase. I nearly fell down it at least three times, so now I knew exactly what it felt like – according to Franco's theory – to wear varifocal glasses. I could even experience the effects without wearing them. After a day of sitting in bed and pottering aimlessly around the house, my headache returned in the evening. The night that followed was a very restless one: my brain, plagued by the stabbing pain, created an incoherent chaos of words and images. It seemed like a superstore had exploded and all the goods were floating along in front of my eyes one by one. My head felt like a lunatic asylum inside. It was filled with a mixture of fireworks, psychedelic madness and a funfair out of control. In the morning, I got out of bed broken and desperate and went to our *medico di fiducia*, *il dottore* Dezza with a heavy heart.

His diagnosis was: *torcicollo*, which had the laconic translation of "stiff neck" in the dictionary. I thought that was quite an understatement considering the tortures I had been through the previous night. Dezza's detailed explanation suggested that my neck had suffered some sort of whiplash trauma. That sounded about right. It seemed that the muscle affected was exposed to an unusually large amount of strain. But what kind of strain? Clearing away the massive pile of wet weeds with the garden fork? The heavy shopping bags which I had to carry from the front gate up four flights of stairs to the top of the house? Or was it a metaphor for Torti being a 'pain in the neck'? Was this an instance of TorTicollo, i.e. Torti-neck? Hopefully it wasn't because there was no known cure for that.

Dezza prescribed me some painkillers. By the end of the day it became clear that they were of no use at all and I had to endure another night from hell. The next day was much like the day before, only now I was taking Dezza's useless pills. By the following morning I was so exhausted by the combined effects of a splitting headache, sleep deprivation and the continuous drilling outside (because they had started stripping off the exterior plaster, yay!) that we decided to pay a visit to First Aid after all. Everything followed the same pattern as when we first came here with our brother-in-law and his torn knee. Nico was not allowed to accompany me to the examination room and spent hours of uncertainty not knowing what they were doing to me (severing my head from my torso? lobotomy?). But in the examination room I was soon put at my ease and was given an anaesthetic infusion. The medics agreed on the diagnosis of *"torcicollo"*. To me it still sounded like quite a euphemism. After an hour's rest and having been given new painkillers, I returned home where I spent another sleepless night tossing and turning with a headache. The painkillers prescribed by the hospital were just as ineffective as those I got from Dezza! The drilling and pounding continued to reverberate across the house all day. The house was like a skull, battered by Torti all day long, while I sat inside barely able to breathe from pain and anxiety. Was I going crazy? Only after a third visit to my *medico* did I receive medicine that finally had some effect: it was a type of painkiller designed to be taken for severe migraine and it was so strong it could knock out a horse. I let out a sigh of relief as I felt the pain subsiding and I could finally lie in bed without my head throbbing.

In the meantime, the exterior walls had been stripped nearly bare of all the plaster, and the red bricks were showing through. I had an X-ray done of my neck a couple of weeks later and it showed that there was a slight curvature in my cervical vertebrae which was causing my head to tilt slightly to one side. Torti was curious to know what was in the big brown envelope which I had brought back from the hospital. I showed him the X-ray. "*È storto,*" he concluded, because he was an expert on X-rays too. In reply I pointed over at the terrace with its newly poured concrete floor and said in Torti's usual tone of voice: "*È anche storto*". Because he had definitely made a hash of that job...

Alle urne

We were late yet again, but the civil servant at the town hall in Montecalvo Versiggia promised to squeeze us in. A little later, we became registered voters without having to pay even a penny. We could vote in two elections: the local elections and the European elections. In our status as residents of Montecalvo Versiggia (because we were in possession of our *residenza*) and as citizens of Europe, we had the inalienable democratic right to vote. And we were going to exercise this right, too, as conscientious, responsible and well-informed citizens. *Alle urne* as the Italians call it. Because the ballot box is ominously called an urn here. We didn't know much about the politics in Montecalvo yet, but we were personally acquainted with both mayoral candidates. One of the candidates was the current mayor, Roberto Delmonte, who once came to our house to give us his pamphlet during the lead-up to the election campaign. We nodded bravely when Roberto enumerated the list of trivial accomplishments he had achieved last year and outlined his future plans for the county. We had our own opinions. The only visible action taken by the council that we had witnessed during our brief year here was the construction of a panoramic viewpoint with a car park near the town hall. The car park was relatively unused, situated on a quiet country road. It had a gravel surface, it was equipped with picnic benches and it was surrounded by a wooden fence. It's true that the view was beautiful, but you could enjoy beautiful views nearly anywhere in this area. Sometime later, the council completed this tourist attraction by adding decorative lights, CCTV cameras and a large mosaic feature. When you drove past it you had to be careful of the large pothole that had appeared last winter at the side of the road just past the *punto panoramico*. The pothole had not been repaired yet because of lack of *soldi*, funding. The government likes to use the smokescreen of having insufficient resources when they can't be bothered to do something.

Delmonte was straying into enemy territory when he campaigned in our *frazione* because his opponent was noone other than our own *antennista di fiducia*, Piero Moro. Piero enquired several times, "You are entitled to vote, aren't you? I am one of the candidates, you know. If I become mayor, everything will change!" He didn't exactly say what would change but probably not the disturbance his dogs caused with their constant barking day and night. A mayor whose job is to protect the general well-being of all the residents yet who doesn't give a damn about his own neighbours (or worse, argues with them vociferously in the street from time to time) did not seem to us to be the right candidate. In fact his actions probably provided grounds for a *'contraddizione in termine'*. We also thought it strange that the second man on Piero's list had a Dutch name, and – as far as we could make out from the locals – this man had never been seen in Montecalvo. Why were there no decent *di fiducia* candidates?

There very nearly was one, but Nando, Leda's husband and the owner of the restaurant Bagarellum, refused to step up. At a conspiracy meeting of the *comitato*, as it was called, to which we were also invited for some unfathomable reason, all the attending members tried to persuade Nando to put himself forward as a candidate. He would make the perfect 'home-grown' mayor! But Nando was careful not to be caught up in the murky waters of county politics. And he was right too, we thought secretly, as outsiders. From what we understood, the *comitato* had been founded some time ago by a number of locals who came from the *frazione* Bagarello and its surrounding hamlets as an organised form of resistance against the headstrong leadership of the civil servants, the mayor included. They wouldn't have shied away from taking legal proceedings if they had to. Their biggest grievance had to do with the road that led in and out of the little hamlets concerned. For starters, it was in a pitiful state because a succession of hard winters made the asphalt surface weaken, break and crumble. And in spite of repeated requests and official complaints, the council had done nothing to mend the road (no *soldi*?). They came up with the excuse that the road didn't belong to the council but that it was a private farm road that had to be maintained by the people who lived in the houses along it. The occupants did not agree; what exactly was the difference between their road and any other road in the county? This exposed another method used by governments to flagrantly shirk their responsibilities: start up a discussion about who is responsible for what.

The *comitato* had another gripe that peeved them no end: the same road gave access to several larger villages and facilities in the area, yet there were long stretches of it which weren't even surfaced. There was a section of a couple of hundred metres which was only gravelled and which turned into an impassable mudbath during the autumn and winter. In the summer, when they drove down it, a layer of dust would cover the locals' carefully polished cars. In this instance, it was the owner of the surrounding vineyards who was standing in the way of road maintenance. According to the council the road was going to stay a farm track as long as this farmer refused to allow it to be surfaced. There was nothing they could do about it. The farmer was continually throwing a spanner in the works, because he even refused to maintain the unmade roads. When you drove past his farm, your car jerking and bouncing, he always gave you a dirty look.

Unfortunately the *comitato* had not achieved much regarding the farmer as yet; the most it could do was to hatch evil plans, something the members derived enormous pleasure from. We joined in the fun. "We should even the road out with a truck-load of sand under the cover of night," said someone. We suggested tarring the road on the sly just in one night. How should we know; we had no idea what form the consequences of such drastic action would take. The atmosphere was cheerful and there was lots of laughter and gossip while the *comitato* members helped themselves to some nibbles and a glass of wine. To us it looked more like a social club where members could give free rein to their crazy ideas. Even so, the *comitato*'s activities had an interesting little upshot when, after much moaning and discontent, one of the locals, Salvatore, managed to get the council to repair two potholes near his house. Unfortunately, his turned out to be a bitter victory because when some time later, Salvatore drove his tractor down a small section of the asphalt road in order to access his vineyard, the mayor was waiting for him in person ready to give him a penalty ticket. It was namely forbidden to drive tractors down asphalt roads. For weeks afterwards, Salvatore would see red every time this little episode was mentioned.

Since Nando refused to stand for mayor, we, as voters, had to make do with the sparse leadership material that we were presented with. A couple of days before the election we received our first real polling card. It was a long card containing rows of boxes to be stamped: one box for each election. We felt we had to repay the trust vested in us and when the big day arrived, we set off to fulfil our democratic duty. Exciting! We wondered how it was going to be organised. We turned up at the polling station and received two voting slips each: one for the European election and one for the local election. But the slips only displayed the logos of the parties involved, and there were a couple of empty lines at the bottom where we would need to write down the names. Which names, was the question? We hadn't learnt the whole list of all the electoral candidates by heart, of course. You could only see their names on posters which were displayed outside the polling station. Luckily, the staff at the polling station let us go outside, where we filled in our voting slips, pressing them against a wall. Out in the open, in full view of nosy passers-by. Voting anonymously in Italy: forget it!

Back inside, Nico posted his voting slips straight into the appropriate ballot boxes, which resulted in immediate fuss behind the officials' desk. "Oi, oi, not so fast, we need to check if your name is on the list!" We were wondering why that was necessary since the council had sent us our polling cards. Moreover, these same officials had given us the voting slips a couple of minutes ago! Maybe this last check was a mere formality? No, Nico was not on the list! Quick as a flash, I posted my votes into the ballot boxes too, before the officials could declare that my name was missing from the list too, which indeed turned out to be the case. Whether we were on the list or not, we had voted, no one could change that now! After some more fuss, the polling station supervisors decided to rectify the illegal voting situation at the end of the election. We decided not to ask how they would do that. Would our votes not count?

A couple of days later, the results of the local elections were displayed on notice boards dotted around the county, the kind which were normally used for death announcements (*alle urne*!) Delmonte, 'the ruling monarch', won three-quarters of the votes, while our neighbour won one quarter. Very clear-cut results. We were curious how our great leader was going to further Montecalvo's interests in the public sphere in the coming years. And we were also interested to see what Piero had in store to make life difficult for the old mayor. Just to be on the safe side, we had agreed that one of us would vote for Delmonte and the other for Piero Moro. In this way if we met one of them on the street we could honestly say that 'we' voted for him. *Salvare capra e cavoli*, to have your cake and eat it: an old custom from our homeland, put to good use in Italian village politics!

La pendenza

One day after work had finished, Torti suddenly announced that in a couple of hours' time, at dusk, he would be coming back to the site with his daughter. It sounded very mysterious: were they planning to put on a cactus exhibition for us? As the sun was going down, we did indeed spot Torti's grey Punto driving up the hill. We couldn't resist the temptation: "*È lui, è lui!*" we cried out. A little later, Torti and his daughter got out of the car, carrying a range of measuring equipment, but there were no cacti in sight. Were they going to take measurements of our plot? The old boss shouted up at us to come down and meet him on the terrace. His hand was making the characteristic calling gesture. Torti's daughter set up a tripod on the terrace floor and turned on a laser beam that was fixed to it. Torti himself, got to work with his indispensable tape measure and carpenter's pencil. In the meantime he explained to us that the terrace had to have a *pendenza*, a slight slope so that the rainwater would run off it instead of collecting in pools against the wall of the house. That would be bad for the *intonaco* and would lead to water getting in. Moreover, the water had to flow towards the west side of the house in order to prevent it from cascading down our drive like a small waterfall. Well thought out. Maybe that potato in his head was not obstructing his thinking as much as we had thought. But it was clear that he had brought his daughter along because she found it easier work the modern laser technology. This didn't stop Torti from bossing her around, although he used slightly more diplomacy than usual.

The father and daughter duo put in half an hour's work, taking all the necessary measurements and making pencil marks on Mimmo's pristine little pillars (all five of them!) until it was completely dark. The last layer of concrete, which would later be tiled over, could now be laid precisely at the right angle. The next couple of days were spent by Mariano and Mimmo spreading the concrete layer under Torti's instructions. When their work was done, it would take a couple of days for the concrete to set and then the tiling could begin. The terrace floor was going to be finished off with an edging made of marble plates, with a slight overhang to keep the water off the plastered wall underneath. After a couple of days' wait, Torti made an enthusiastic start on laying down the marble plate edging. First the cement layer around the edges had to be adjusted to accommodate the difference in thickness between the marble plates and the tiles. Torti kept a close eye on the long and labour-intensive job of chiselling away just the right amount of concrete to achieve a level floor. From the abrupt shouting and cursing that suddenly broke out, we concluded that things were not going well. Torti's screaming was a sure sign the *pendenza* was not right. "*È sbagliata, è sbagliata,*" he shouted. It was a disaster. Despite employing the high-tech instrument and the expertise of Torti's daughter, the concrete did not slope in the right direction. Now what?

There was no other option but to break up a large part of the concrete floor. This was a perfect job for poor Festi, the untrained labourer, who was also given the unpleasant job of adjusting the basement stairwell when the staircase didn't fit. He didn't seem very pleased with this task, probably because Torti supervised him closely and told him exactly how much had to be taken off every square inch of concrete. Luckily it all worked out in the end and they were able to finish fitting the marble edging once they had waited long enough for the new layer of concrete to set.

Luckily, in the meantime, there were plenty of other little jobs to get on with. For example, Torti had to prepare the driveway (which extended partly underneath the terrace) for filling with concrete all in one go. Torti's tape measure came out again and was put to good use. He left the laser measuring device at home this time. Even without the interference of modern technology, Torti found enough to mutter and complain about. He tried several times to tell us something about *pendenze* and other nuisances but we didn't understand what he was going on about. It all finally became clear when Cassani next visited and provided us with a patient, clear explanation, using the piles of bricks that Torti had left here and there, to illustrate. The piles indicated the required height that the concrete would have to come up to, taking into account the desired direction of the water flow. At the spot where our new entrance would be, there was a little hill in the driveway. This meant that at this spot the layer of new concrete would only be about ten centimetres deep. This was thought to be undesirably thin because it would be prone to crumble easily. One option was to raise the level of the whole driveway which would cause more problems because it would place our driveway too far above street level and make it hard to bridge the gap. It was a design problem and Cassani would need to work out possible solutions at the office in discussion with his colleagues. Finally, the cement lorry would need to pour the concrete in exactly as prescribed. Our driveway was going to become a work of art!

The creation of our new cement driveway was indeed quite an event. First of all, Torti ordered a proper cement truck to deliver the cement, one with a proper cement mixer on top. Marco had often complained in the past that Torti was always too stingy to have the cement delivered, so they always had to mix the cement by hand, the old fashioned way, using a simple concrete mixer. They did this for every building project, for the long boundary wall, the large terrace floor, for the wall at the bottom of the slope near the car park, etc. Hundreds of kilograms of sand and concrete had passed through their hands.

The cement truck was escorted by a real *gru*, which was quite a bit larger than the rickety little crane on the back of Torti's 'Dumpy'. The crane was needed to guide the discharge chute, which was weighed down with concrete, into the correct position. The concrete finisher inspected all the directions about what depth of concrete was required, and where, and dauntlessly carried on with his task. He only had one attempt to get it right: stopping mid-way would result in a break-line in the driveway.

We watched in awe the expertise with which the man guided the discharge chute; he had to manoeuvre it around the terrace in order to cover the whole area of the driveway. The job went swimmingly and within an hour the drive was covered in a smooth layer of cement. This was quite a change from all the various stages that the drive had undergone in the past couple of months. First it was nothing but a rough layer of concrete with bits of plastic poking out here and there. Then it was the same again, only now broken up along the boundary line. Then it was the same rough concrete drive but now with a new wall built along the boundary line and with rubble, sand and concrete everywhere. Then the same again but now complete with grooves and drainpipes. Then it got covered in marks left behind by the supporting beams used during the building of the terrace, and so on... Finally, it was transformed into a splendid, pure, smooth, beautiful surface. It acted like a soothing balm for our tort(i)ured souls. We could park our car in front of our own house on our own drive... at least if Torti's 'Dumpy' didn't need the space and if we didn't need to clear the way for his diggers.

But whichever way we looked at it, there was no denying it: the end of Torti's *grande lavoro* was in sight. For some of his labourers it was literally their last *grande lavoro* with Torti. Mimmo, the small, friendly Sicilian, had decided to quit and had been working on a different building site for the past couple of days. He was fed up with Torti's abuse: "I am a trained *muratore,* I know what I am doing. All Torti needs to do is tell me what needs to be done and check it when it's ready. It makes my blood boil when he stands there, breathing down my neck and criticising my work!" Mariano, the very competent and patient Moroccan *muratore*, also seemed to have had enough of Torti's dictatorial management style and conflicting instructions. In the same vein, Torti asked Mariano this morning to warn the neighbours that the cement mixer would be blocking the entrance road for a couple of hours, yet when Mariano wanted to warn the neighbours about this the day before, he had stopped him because he didn't think the road would be blocked at all. And the day before that, Torti had given him instructions that he *should* warn the neighbours! Another time he told Mariano to ring the tiler ("*Subito!*") to ask him to come and start laying the tiles on the terrace the next day because there was rain expected the day after that. The tiler was surprised to receive Mariano's call because he had spoken to Torti over the phone the night before. Was the old man slowly losing his marbles? Maybe his mental fitness had a slight *pendenza* of its own.

Il corso di cucina

"*Sei un dentista,*" said Nando as he watched me pressing the pasta dough into the metal *ravioli* mould. With my long, clumsy fingers, I was carefully pressing the dough into the corners of each compartment one by one in order to prevent the little pasta pillows from falling apart when I tapped them out of the form. But my precise and meticulous approach was obviously not the way it was supposed to be done. Nando's remark made me laugh out loud because the *ravioli* mould really did look like a row of hollow teeth cavities which we had just filled with a mixture of herbs and *ricotta*.

Leda, the real chef, showed me how she did it. She laid the fresh sheet of pasta over the top of the mould and patted it firmly with the flat of her hand a couple of times to bind the two layers together. You needed brute force, not precision, to make good *ravioli*! You shouldn't be too careful either when kneading bread dough, said Leda. You are supposed to abuse it ("Just think of your partner whilst kneading!"), roll it forcefully and pull it apart to break up the gluten. The kitchen is like a torture chamber. Nando always kept his distance when Leda was kneading dough, he said, because her arms were as strong as a rower's after all that demanding work in the kitchen. "*È pericolosa*! She is dangerous!" One punch from her could land you in hospital.

Watching Jamie Oliver's cookery programmes kindled our enthusiasm for Italian cuisine, and a couple of years ago we had bought an Italian pasta maker, to try making the most intricate of pasta dishes: the delicious, filled pasta pillows, or *ravioli* in Italian. Our first attempt was doomed to fail because the dough was too thin and water seeped into the pillows during the cooking process. After this experience we stuck resolutely to making only *lasagne* and *tagliatelle*. But now that we lived in Italy we were given the chance to learn how to make real *ravioli* and not just from anyone but from star chef Leda herself. We grabbed the opportunity with both (left) hands. Some time ago, Leda had told us that she would love to give cooking lessons to our future guests and today's lesson was a test run. And of course, an opportunity to acquire a couple of fine techniques (or forceful ones), now that we were here anyway.

Leda admitted that she had been up all night with a dictionary, pen and paper to learn the most important English cooking terms by heart. *aglio = garlic, cipolle = onions, melanzana = aubergine, cucinare = to cook*. Repeat: *aglio = garlic, cipolle = onions, melanzana = aubergine, cucinare = to cook*. It would be useful to know some English because most of our future cookery course participants would not be able to speak or understand Italian. Leda showed us her A4 sheet of paper containing a long list of cooking terms from Italian to English, which she had looked up in her ancient school *dizionario*. At school, she was good at English, but her knowledge had faded over the years, she told us. She laughed nervously at her 'homework' as she stared hopelessly at the long list of unpronounceable English terms. "*O, signor, signor,*" she said with a sigh, with her gold-rimmed reading glasses pushed up on her forehead. This was followed by another nervous giggle. Compared to the language challenge, the preparation of the traditional Italian dishes was a piece of cake, all the more so because the recipes had been handed down through generations in Leda's family. And wasn't Italian *the* one and only language spoken in the kitchen after all? And now it had to be done in such a strange language. Well, we would see about that.

Once Leda became absorbed in the preparation of all the different dishes, she soon forgot about the English aspect of the course. The list of terms lay abandoned in a corner, partly covered in flour. Leda chatted on cheerfully in Italian, using lots of Italian hand gestures and made herself understood remarkably well. Once in a while, the flow of Italian was interrupted by a random English expression, pronounced in a heavy Italian accent. Leda's bubbly laughter compensated for the language barrier and the workshop concluded without problems. She would never use a pasta mould to make *ravioli* when she did it herself: she would prepare it like a real Italian *mamma*, by hand. Making *ravioli* pillows by hand was a lot more fun and it looked very easy. First, you had to put the pasta machine on the thinnest setting and make two sheets of fresh pasta. Once those were ready, you would use a piping bag to squeeze a row of little heaps of filling on the first sheet of pasta (this would become the bottom layer). This was a tricky task because you had to be careful not to make the little heaps to small or too big and the individual piles had to be spread at just the right distance apart. Next, you had to line up the second sheet of pasta exactly over the top of the bottom layer and use quick karate chops (violence!) aimed exactly between the little heaps of filling to stick both layers of pasta together. Once all the edges were securely pressed together it was time to use the little round *ravioli*-stamp. You had to position the stamp over a little pile of filling and press down and twist at the same time in order to get the perfect *ravioli* shape. Cook the pasta in two minutes (because it was fresh!), prepare a quick pasta sauce, mix it in and you're done.

What else did Leda show us? She demonstrated how to make plum cake using 'tortured' dough and there was also a quiche which had *silene* as one of its ingredients. She picked the *silene* in the meadows and it looked much like a wild flower called red campion. Sometimes she used nettles or *erbette*, and all kinds of weed growing between the paving slabs, as ingredients. Leda's cooking always relies on seasonal produce from her own *orto*, or kitchen garden, and the fields. Nando and Leda keep their own chickens for eggs and meat, and they know hunters for when they need to order game. In the autumn they regularly go out to collect wild mushrooms. Our first workshop had a very fitting finale: a lovely meal shared with Leda, Nando and their son Marco. Our glasses were filled and refilled again with Bagarellum's sparkling house wine, and after a marathon of culinary indulgence we ambled back home for a well-earned siesta. We were sure that the cooking workshop would become one of our top attractions! And if the language barrier proved to be an obstacle, we would simply accompany our guests to the workshop. As interpreters.

L'intonaco

"*Non sei capace,*" said Torti as he watched me walking purposefully towards the scaffolding, holding a bucket in one hand and a paintbrush in the other. "You can't do that." Ouch, one of my legs just started aching. The mocking expression with which Torti uttered the words did not lighten my mood. Why couldn't I do it? How would he know? He hadn't even seen how I was going to do it yet! Anyhow, first we were only going to apply the *isolante*, the transparent primer, onto the brand new plaster surface of the exterior walls. We couldn't go wrong. You couldn't see any of the runs, drips or smudges left by the brush strokes and the primer would be covered up by a further two layers of paint anyway. What was he on about, the old fusspot? Such thoughts were running through my mind but I held my tongue, pretended that I didn't hear him, climbed onto the scaffolding and started painting. Torti went quiet and sauntered off. Some time later I heard his voice bellowing from the direction of the garden where his last loyal labourers were doing groundwork, clearing the rubble, levelling and digging.

After the week of my *torcicollo,* during which the exterior of the house had been stripped bare, Mariano, with the help of Marco and Festi, had replastered all the walls in record time, leaving them with a new smooth finish. Clad in its fresh grey coat, our house was already a lot easier on the eye than when it was covered in the dirty old rough, crumbly and cracked layer of plaster. But we wanted to dress our house up in a pretty, cheerful colour, as was the custom in the region. Yellow is of course always a favourite, closely followed by orange, the latter being our colour of choice: orange would stand out nicely against the deep green foliage of the vineyards behind it. Our villa, a juicy little orange in the midst of all the green! It was quite a task to find a shade that was not too light yet not too dark either: too light and the strong summer sunlight would soon bleach it to white; too dark and the house would look gloomy.

The paint specialist in Broni, recommended to us by Torti, showed us some samples. The choice was easy: we both picked out the same shade of orange. We thought it would go well with the grey damp proof course along the bottom of the house and the grey borders around the windows. We couldn't wait to finish our house in these lovely new colours, but before the painting could begin, we first had to apply the primer, the *isolante*. This meant days of clambering on, over, under and across the scaffolding surrounding the house. And as inexperienced builders, we had to be careful not to stumble or take a step backwards into one of the openings and fall all the way down. Because our villa was not wheelchair friendly and the DURC only covered Torti's employees.

We were working quietly, *nascosto*, so as not to attract Torti's attention, because as soon as he saw us he always had something to say. *"Non cominciare, è ancora bagnato!* Don't start yet, the wall is still wet!" he shouted, for example, because he believed that you could only start painting when the *isolante* had completely dried. But Mariano, who is a qualified decorator, told us quite the opposite. You were supposed to paint over the waterproofing primer whilst it was still fresh. While Torti was restlessly pacing up and down at the bottom of the scaffolding, we saw Mariano standing behind him with a look of bemusement. Another time Torti told us his opinion about our choice of grey paint. *"Non è bello.*

Sembra blu. That's not a nice colour, it's nearly blue." But what pushed Torti over the edge was when I began painting a self-designed decorative border around the window frames. Apart from the grey damp proof course along the bottom of the walls and a couple of grey windowsills, there was nothing to break up the 'orangeness' of our villa. In order to prevent it from turning into one great orange cube, I decided to paint a grey border above the windows, inspired by the elaborate plaster reliefs often seen on real villas. Our humble abode would of course have to do without real plaster reliefs, but we could have a budget version: a kind of *arte povera*, 3D street art. The top borders above the windows were the same width as the windowsills but they were slightly longer on either side and tapered down towards the ends, creating the illusion of 'depth'. Moreover, the variation in the width of the borders lent them an elegant finish. I thought.

Torti immediately noticed my preparations and couldn't resist commenting: "*Che cosa stai facendo*? What are you doing?" I didn't feel like getting involved in a discussion and gave a short, evasive answer. Torti ambled away but kept a close watch. And indeed, I had hardly even finished the border above the first window before he appeared out of nowhere and shouted up at me in a loud commando voice: "*Non è bello, non si fa così.* It doesn't look right, that's not how you should do it." Next to me on the scaffolding there was a full bucket of grey paint and I needed to control myself not to tip it over, turning Torti into a pretty *blu* Italian smurf. The whole idea of a border above the windows was wrong and the fact that the border tapered off at both ends creating a 3D effect was pure deception and completely unforgivable. Italians like their traditions and this was unheard of. The Dutch, on the other hand, are rather a stubborn sort, so the blueish grey borders stayed!

Luckily, we managed to make ourselves virtually invisible and made great progress with the painting and decorating. It was very rewarding work because we saw the house come to life: it really blossomed! Putting the finishing touches to the house also meant that the end of the gruelling building project was in sight! As if that were not reason enough to cheer up!

Degustazione guidata

The entrance to the farmstead was quite impressive; its gate was decorated with a coat of arms and images from mythology. It made such an impression on us that even though we had driven past it several times we never actually dared to go in. *'Azienda Agricola Travaglino, dal 1868 grande vini dell'Oltrepò Pavese'* said a big sign above the entrance. That sounded very attractive to not entirely abstinent people such as ourselves. We wondered what alcoholic treasures might be hiding behind the walls of this old wine farm. Yet we were reluctant to enter; we were worried that the smooth salesmen would immediately seize their chance to sell us a couple of super-expensive crates of wine. The Dutch are famous for being prudent and thrifty, characteristics that in Italy are mainly ascribed to the people of Genoa.

Luckily the wine maker organised an open day, giving us an opportunity for a visit. That meant that we could nose around in relative anonymity, lost in the large crowd of wine fans who would arrive to taste the produce of the wine farm. We had to take this chance, it was now or never. Moreover, the afternoon programme also contained a traditional 'cheese rolling' competition and that was something we wouldn't have wanted to miss. On the appointed Sunday of the open day, we set out and parked our car at the top of the road so that we could walk through the wine producer's vineyards. Every vineyard plot had a little sign depicting the bottle and the type of wine that the grapes in that plot produced. There were a variety of wine types, described by imaginative names such as *Campo della Fojada* (Riesling), *Poggio della Buttinera* (Pinot Noir), etc.

The wine farm was in a beautiful location, on the hillside, its buildings surrounded by woodland, which we were later told was home to some deer. When we arrived in the courtyard we were proved right: there was a good-sized crowd. Exactly the right way to satisfy our curiosity without being seen or feeling under pressure. To start with, we nosed around the grounds, looking at the machines and piles of stock, before entering the *enoteca*, the shop of the wine farmer. We foraged around among the pretty bottles and gift boxes on our way to the tasting room. Once there, we settled on the stools around one of the many huge wine barrels. Fortunately we weren't on our own here either, we didn't feel lost or abandoned. Nearly all the barrels were completely occupied and Travaglino employees were walking up and down the rows, serving wine as well as bread and small blocks of Parmesan. Because we all know that you shouldn't drink on an empty stomach. Every table had a spittoon on it but we were not planning to use it! The Dutch Genovese were not going to let expensive wine go to waste by spitting it out.

After some time we were approached by a female wine connoisseur. Would we like to try some wine? We nodded greedily (and thirstily). We explained to her in slightly incoherent Italian, that since we knew nothing about wine, we would be very grateful for some advice. We knew that this was a careless revelation because now we could easily fall prey to the wine merchants. Our expert advised us on the order in which the delectable wines had to be tasted (we tried five kinds in the end). She was wise enough to make sure we would finish up with the most expensive and most delicious wine. It was a red wine that went under the name of *Buttinera* Pinot Nero. It was a bit on the pricey side because its manufacturing involved resting it for a couple of years in an oak barrel. Well, it was expensive in comparison to the supermarket wines we drank in the Netherlands costing 4 euros a bottle, and here in Italy we had never spent more than two and a half euros on a bottle of wine in the Cantina Sociale. The wine tasting session ended with a tour of the winery itself. Apart from the several-metre high aluminium containers where the wine was stored, we were most impressed by the historic wine cellar with its vaulted ceiling, dating from the year 1111 (!) and its endless rows of *barriques*, oak barrels.

There was no escape: we had to take home a case of delicious *Buttinera* (thirteen euros a bottle) and another one of the spicy Riesling (nine euros a bottle). We wanted to carry on with the tasting at home on the sofa. It was the perfect arrangement: after enjoying a couple of glasses of wine you can't feel the pain of parting with your money. We were about to get the car to load our alcoholic treasures into it and leave. But wait, it was time for cheese rolling!

And indeed, the gladiators were already lining up on the road, outside the farmstead. It turned out to be the national cheese rolling competition. This was a serious sport as you could see from the 'athletic' build of some of the competitors. They looked like they ate the cheese up after the race. It was fun to watch; it looked like a variation on the Dutch game of *klootschieten* or 'ball shooting', practised in Eastern Friesland. Every competitor tried to make their cheese roll as far as possible and to finish the set distance in the least possible throws. Some of the cheeses burst open on the way and Saar took full advantage of this as she rushed to hoover up the tasty treat much to the amusement of the crowd.

We had completed our first ever wine tasting experience in the Oltrepò. Just a couple of hundred *cantinas*, cellars, left to visit. *Salute, cin cin* or *cent'anni*, as they say in Italy!

Il primo tuffo

The tap had been on pretty much the whole day and the water level in the swimming pool was so far about waist high. In the meantime, Torti and company carried on digging around the pool and creating a poolside area. Torti was working the digger, Marco and Mariano were using the spades and the wheelbarrows. After weeks of waiting for the winter groundwater to go down and a couple of days' worth of contemplating and taking measurements, the Dutch flat-pack swimming pool was finally built. There was nothing else for it but to take the first *tuffo*, the first jump! It involved a bit of climbing because the steps hadn't been fitted yet (and nor were the pump or the filter installed) and the water was ice-cold, but it was ready! And with this first jump the Villa I Due Padroni swimming pool was officially opened. At the cost of some frozen body parts.

The installation of the swimming pool we had bought in the Dutch sales was a lot more difficult than we had expected, even though the swimming pool specialist swore it would be child's play and it would only take two people to do. A local specialist (*piscinista*?) even advised us against digging the swimming pool into the ground: the pool's steel folding walls were not designed for that. But of course, was he a *di fiducia* expert? Or was he just trying to flog his own merchandise to us? Our swimming pool was designed expressly to be dug in. Wasn't it? Unfortunately Cassani and Torti agreed that the pool shouldn't be dug in. We had already determined that, every winter, the ground water came up nearly to surface level where we live (a fact that caused great astonishment to drainage expert Cassani) and that at the same time of year, the roads buckled under the weight of the water pouring down the hillside: it was common to see splits in the asphalt, sometimes leading to full-blown *frane*, landslides. If we dug a swimming pool into the ground here, we would risk it being pushed up by the groundwater in the winter, or it might slide down-hill towards the valley.

What was there to do? The solution was simple and one that pleased Torti no end: we had to build walls to stop the soil above the pool from sliding downhill, and we had to construct a foundation made of reinforced concrete which would resist the pressure of groundwater. At least two walls: a vertical one and a horizontal one. Torti could hardly contain his happiness, even more so because this work was not included in the building plans, which meant working overtime. And we had just enjoyed a wonderful Torti-free month! But we couldn't afford to delay the construction of the swimming pool any longer because the first official guests of the Villa I Due Padroni would be arriving soon and they were undoubtedly expecting a lovely swimming pool facility. We didn't know yet that some guests had very different priorities when it came to sport. Torti started digging and pouring cement and Nico studied the pool building instruction booklet. The instructions were very complicated and Nico wanted to be well-prepared to deal with Torti's stubborn objections. Nico put his faith in Mariano's intelligence because he knew they could work well together. But it was of utmost importance not to let Torti stick his oar in.

Nico and Mariano did a good job at keeping Torti at bay. Once the cement foundation was poured and set, a frame was built on it which would keep the steel walls in place. It was mainly a question of good preparation and taking precise measurements. One of the most exciting parts was fitting the liner: it had to, as it were, fall into place as we were filling the pool with water. We were worried that it would tear. What would be the use of a leaky swimming pool? Luckily the liner didn't tear, but it didn't manoeuvre itself correctly into place at the first attempt either. We had to drain the water by hand before we could make a second attempt. This was successful and we could carry on filling the pool. That would take some time because thirty cubic metres is quite a lot of water. The only funny thing was that now that we had built the walls and laid the foundation, we had practically created a complete concrete swimming pool. After building one more wall and some tiling it was finished. Our Dutch swimming pool bargain turned out to cost us a lot more than we had hoped.

Finishing the swimming pool had finally concluded the building project, *il grande lavoro*. The terrace was ready, the house was painted, our car was on the drive, and we could take a dip in the pool. We no longer had to worry about how to invest our money. We could carry on with painting the railings, planting the garden, and furnishing the terrace, without further interference from obstinate contractors. But by now, the daytime temperature outside was in the thirties and the new deckchairs and the cool water were looking very tempting. The Italians didn't invent the *siesta* for nothing, and after the *siesta* it's really time for an *aperitivo*. I don't know, those Italian Dutchmen and their idea of hard work...

Epilogue

We hadn't clapped eyes on the first guests staying in our brand new villa for nearly the entire week that they were there. In the mornings they would go out for a quick trip in the car, never for long, apparently just to do some shopping. They would rarely leave the house. They sat indoors every afternoon with the shutters closed even though the weather outside was beautiful. What was going on? We welcomed them with some complimentary wine and gave them lots of tips for days out. In the living room there were a couple of thick folders full of tourist information. Were the guests not enjoying themselves in our villa despite all our efforts? We sat nervously biting our nails every evening. We didn't get it.

On the Saturday of their departure we entered the apartment with a heavy heart, expecting critical comments that we would need to take on board. But to our great surprise the guests welcomed us with great enthusiasm! "It was awesome!" they exclaimed in unison. "We could watch Eurosport on your TV and we were glued to the Tour de France live broadcast, just like at home!"

You've reached the end of my book; I hope you enjoyed it!

If that is the case, **would you PLEASE post a REVIEW** on Amazon (or Goodreads or any other site you are acquainted with)?

As a self-published author I can use every review I can get!

Thanks!

Glossary

a norma – according to regulations
a offerta – voluntary
a posto – OK
a voce – read out loud
acquedotto – water supply
affare – bargain
agenzia immobiliare – estate agent
aglio – garlic
agriturismo – farm holiday let
all'improvviso – unexpectedly
al dente – just cooked
alle urne – to the polls
amico – friend
andata e ritorno – there and back
antennista – TV aerial installer
antipasti – starter
aperitivo – aperitif
architetta – architect
argento – silver
arte povera – 'poor art' style
articolate – folding, on hinges
asse – toilet seat
assegno – cheque
astemio – teetotaller
autista – chauffeur
autonomo – independent
bagnato – soaked, wet
bagno alla turca – squat toilet
barista – barman
barrique – oak wine barrel
bastardo – bastard
bloccato – locked up (knee)
borlotti – a kind of bean
bozza – draft
brutto – ugly
buffone – joker
buongustaio – foodie, gourmet
buonissimo – very good
cancello – gate, fence
cantina – basement, wine cellar
cantina sociale – wine cooperative
canzone – song
caparra – deposit
capo pigro – lazy boss (Torti)
capriolo – roe deer
casalinga – home-made
cent'anni – cheers
chimica – chemical adhesive
ciabatta – slipper
cin cin – cheers
cipolla – onion
cipollata – onion festival
clandestini – illegal immigrant
codice fiscale – national tax nr
cognome – surname
collaudo – the MOT
colonello – colonel
colpa sua – his fault
come mai – how come
comitato – committee
commercialista – accountant
complimenti – compliments

compromesso – prelim. contract
computo metrico – cost estimate
condominio – apartment-complex
contatore – meter
conto corrente – current account
contraddizione – contradiction
coppa – cured ham
coppia di fatto – partnership
copritetti – roofer
così così – so so
crostata – fruit tart, pie
cucinare – to cook
dentista – dentist
di fiducia – trustworthy
di una volta – as in the old days
digestivo – after dinner liqueur
dilettanti – amateurs
disegno – design
dizionario – dictionary
dolci – desserts
domani – tomorrow
donazione – gift, present
donne – ladies
dottorandi – PhD students
dottoressa – doctorate
è lui – it's him
è piccolo il mondo – small world
elettrodomestici – appliances
elicottero – helicopter or machine
emozione – emotion
enoteca – wine shop
erbette – a type of edible weed
espresso – espresso coffee
extracomunitari – non EU people

falegname – carpenter
fannullone – fool
fare bella figura – look good
farmacia – pharmacy
farmacista – chemist
fatta in casa – home-made
frane – landslide
fratelli – brothers
frazione – hamlet
frittata – omelette
funghi porcini – wild mushroom
fuori – crazy, 'out of it'
gabinetto alla turca – squat toilet
geometra – architectural engineer
ginocchio – knee
giornalista – journalist
gnocchi – soft dough dumpling
gnocco fritto – fried dumpling
gommista – tyre salesman
grande lavoro – building project
grappa – alcoholic grape spirit
gru – crane
idraulico – plumber
il bagno – the toilet
il water – toilet bowl
immobiliare – real-estate
impari – odd, unequal
importazione – importation
imposta di bollo – tax stamp
in bollo – level
in collina – in the hills
in omaggio – for free
in ordine - OK
infortuno – accident, misfortune

ingegnere – engineer
interruzione – interruption
intonaco – plaster
isolante – primer
macchina – car
laboratorio – workshop
lavanderia – launderette
luce – light, electricity
macchina – car
mafia – mafia
mal di testa – headache
maleducato – ill-mannered, rude
malumore – bad temper
marmi – marble plates
medico – doctor
melanzana – aubergine
meno male – thank goodness
meridiano – southerner
meteo – the weather
mi dica – go on, can I help you
moca – percolator, coffe pot
modulo – form
monosillabi – monosyllables
muratore – bricklayer
nascosto – secretly
nocino – walnut liqueur
nome – first name
non c'è male – I'm OK
non ci sono problemi – no problem
non sei capace – you can't do it
non sta bene – is ill
non trasferibile – non-transferable
nonna – grandma
notaio – solicitor, notary

occupato – occupied
ora solare – winter time
ora legale – daylight saving time
orata – sea bream fish
orto – kitchen garden
orzo – a drink made with barley
padrini – godfathers
padroni – masters
pagine gialle – the Yellow Pages
palazzo – palace
pancetta – pancetta bacon
passeggiata – a walk
passione – passion
pattini – skates
pavimento – floor
pazzo – crazy
pendenza – slope
pericoloso – dangerous
perizia – technical testing
permaloso – prickly, touchy
permesso – permitted
persiane – shutters
persone brave – honest people
persone serie – serious people
piano piano – slowly, take it easy
pico bello – perfectly done
polenta – polenta, maize flour
ponte – long weekend, bridge
porca miseria – what a disaster
potente – potent, powerful
poverino – poor creature
pranzo – lunch
prenotazione – reservation
prestito – borrowing
presto – quick
prete – priest

preventivo – quotation
primi – main courses
la professoressa – lecturer (female)
progetto – project
pronto – yes, what's up
pronto soccorso – First Aid
proposta d' acquisto – offer to buy
proprietà privata – private prop.
prosciutto – ham
prosecco – sparkling wine
puntuale – punctual
quantità – quantity
radicchio – Italian leaf chicory
raffreddore – cold (illness)
ravioli – ravioli
razza maledetta – cursed race
residenza – residency
ricorsa – modification
riduttore - regulator
ringhiera – railing
riscaldamento – heating
risotto – risotto
robinia – locust tree
rompere i coglioni – to annoy
rustico – brick shed
salame – cured sausage
salami – salami (cold meat cuts)
salotto – living space
salute – health
salvavita – circuit breaker
santuario – temple
sbagliato – wrong
sconto – discount
secondi – second courses (meal)

segnalazione – signs
signore – ladies
signori – gentlemen
silene – red campion (wild flower)
soldi – money
sommelier – wine connoisseur
sopralluogo – on-site inspection
spettacolo pirotecnico – fireworks
sportello – counter
spumante – sparkling wine
stanza – room
stasi – stalling
storto – askew, lop-sided
strollato – cement layer
stufa – (woodburning) stove
subito – immediately
tabaccheria – tobacconist
taglia media – medium-sized
tartufo – truffle
telecomando – remote control
tessera sanitaria - health certificate
torcicollo – stiff neck
torrente – creek, stream
trave – supporting beam
tubi – pipes
tuffo – dive
turbativa – disruption
tutto bene? – everything OK?
veline – eye candy (on TV)
vergognati – shame on you!
vertigini – dizziness
vieni – come here

Made in United States
Troutdale, OR
07/26/2024